Slaying th
Vaccine Dragon

Judy Wilyman, Saeed Qureshi, John O'Sullivan and Robert Beatty

Slaying the Virus and Vaccine Dragon

Print ISBN 978-1-949267-98-3
ebook ISBN 978-1-949267-99-0

Other books by The Dragon Slayers:

The Sky Dragon Slayers—Victory Lap
Slaying the Sky Dragon: Death of the Greenhouse Gas Theory
The Deliberate Corruption of Climate Science—Tim Ball
Human-Caused Global Warming—Tim Ball
Buffoon: One Man's Cheerful Interaction with the Harbingers of Global Warming Doom—Ken Coffman
Black Dragon: Breaking the Frizzle Frazzle of THE BIG LIE of Climate Change Science—Geraint Hughes

STAIRWAY PRESS—Apache Junction

Cover Design by Guy Corp
www.GrafixCorp.com

STAIRWAY PRESS
www.StairwayPress.com
1000 West Apache Trail, Suite 126
Apache Junction, AZ 85120

In Memoriam

To the millions of people who had their health, livelihoods, and family associations—and lives—affected by the COVID-19 phenomenon.

The affects are associated with 'vaccines' promoted by the World Health Organisation, national authorities, many media outlets, various commercial institutions, and several levels of governments.

Of particular concern are the mandates associated with employment, and our education facilities. Results from these intrusions caused significant breakdowns in societies around the world, which are being experienced by current, and likely to be by future generations.

This book is dedicated to our friend, colleague and mentor, Dr Tim Ball. Requiescat in Pace.

Foreword

FIRSTLY, WHY WOULD the authors choose me to write the foreword to this book? That is a particularly relevant and important question to ask. In this Foreword, I will attempt an answer.

Prior to the current COVID-19 pandemic, I lived a very comfortable life in retirement following a long professional career working within the Australian pharmaceutical industry either as a member of staff, and later, as a senior consultant in clinical trials and drug regulatory affairs. I started and owned Australia's first full-service Contract Research Organisation where I employed doctors, nurses, pharmacists, statisticians, scientists and health economists. Over many years I consulted for more than half of all the multinational pharmaceutical companies—I was a Big Pharma guy.

But then COVID came along and there were certain things that unsettled me about the gene-based so-called COVID "vaccines" and then about what was being said about ivermectin. I knew it was not possible to develop a completely new generation of gene-based "vaccines" in 10 months under Operation Warp Speed and I also knew the public was being told lies about the safety of ivermectin which is an important, relatively safe and essential therapeutic agent used worldwide.

I began to spend time researching the mRNA COVID-19

"vaccines" and the more I learned, the more concerned I became. I had no previous strong concerns about vaccination in general. I was vaccinated as a kid and my kids were vaccinated according to the usual vaccination schedules. But something was wrong—very wrong.

During the pandemic, I was blessed with another grandchild and became aware of the incredible number of childhood vaccinations now recommended for children. This added to my concern. Why would a newborn need a hepatitis vaccine? What was the evidence? What was the safety? Vaccines should not be assumed to be safe drugs. Many have been withdrawn due to death and injury but the public is not generally aware of this.

The speed of development of the COVID-19 "vaccines," the extremely limited nature of the clinical safety and efficacy data submitted to the Australian drug regulator (the Therapeutic Goods Administration, TGA) and the speed of Provisional Approval of this new class of gene-based "vaccines" was alarming. The alarm bells were ringing.

Since then, everything the so-called "health experts" said about the COVID-19 so-called "vaccines" was wrong: they were not safe and they were not effective. These "vaccines" did not prevent infection nor did they prevent transmission of infection. In fact, the data now clearly shows these so-called "vaccines" (in actual fact they are not really vaccines in the usual sense of the word because they do not prevent infection) produced more recorded deaths and serious adverse events than any drug released in the history of the pharmaceutical industry and they have only been on the market for a little over 2 years.

I only learned late in 2022 that the COVID-19 so-called "vaccines" were developed under the complete command and control of the US Department of Defense and under US legislation they are properly termed "countermeasures." As such, the usual development and testing guidelines for quality, safety and efficacy used by the pharmaceutical industry were not required and, in many cases, were simply not done. This is why the COVID-19 so-called "vaccines" have proven to be so unsafe and should be withdrawn immediately from the market. It usually takes about 10 years to properly develop and test a new vaccine. Even then, there are many examples where conventional vaccines were originally thought to be safe but were later found to suffer from serious safety issues and withdrawn from the market.

In the last 20 years , more than 100 approved drugs were withdrawn from the market due to safety concerns. It normally takes several years for the true magnitude of the safety concerns to be fully recognised following drug marketing approval. Nobody knows the long-term effects of the gene-based COVID-19 "vaccines." The total damage due to these experimental "vaccines" may be far worse than anyone imagined. Only time will tell the true dimension of the problem.

Since the introduction of the COVID-19 "vaccines" there has been a significant increase in unexplained deaths around the world due to heart attacks, strokes, cancer and neurological conditions. No government has provided a satisfactory explanation for this phenomena and it appears the COVID-19 "vaccines" are the most likely cause.

Despite the known human toll taken by these gene-based COVID-19 so-called "vaccines" and despite the fact that none of them have been fully approved anywhere, governments are pressing ahead under confidential public-private-partnership arrangements to build an even larger vaccine industry based on mRNA "vaccines" even though there is no long-term safety data.

The catalogue of misinformation, disinformation and outright lies told by the vaccine manufacturers and the complicit behaviour of drug regulatory agencies in many countries has rocked my world. It has been a difficult awakening for me personally and it has led me to a general and healthy distrust of the claims made by the vaccine industry. I ask myself, if they have lied so repeatedly and blatantly in relation to COVID-19 "vaccines"—what else have they been lying about?

If there is a single lesson that emerged from this COVID pandemic, it is the simple recognition that you cannot trust Big Pharma to tell the truth; you cannot rely upon the advice of so-called "health experts" connected to Big Pharma, either directly or indirectly, to act in your best interest. You must question everything.

That is what this book is about and that is why this book is so important.

—Phillip M. Altman
BPharm(Hons), MSc, PhD
Clinical Trial & Drug Regulatory Affairs Consultant

Phillip has a Bachelor of Pharmacy (Hons), a Bachelor and Masters of Science and a Doctor of Philosophy. He works as a clinical trial and regulatory affairs pharmaceutical industry consultant with more than 40 years of experience in designing, managing and reporting clinical trials. Dr Altman has dealt extensively with the Australian Therapeutic Goods Administration throughout his career.

Dr Altman has worked for and consulted with most of the international pharmaceuticals represented in Australia. He was fundamental in the establishment of the Australian Regulatory and Clinical Scientists Association (ARCS), which is a peak educational forum for more than 2,000 clinical and regulatory scientists working within the Australian pharmaceutical industry. He has a Life Membership of this Association.

Phillip's experience involves more than 100 clinical trials (covering Phases I,II, III and IV, i.e. from first administration to animals, then man, to post-approval trials), and a similar number of new drug applications, TGA appeals, and applications to modify existing approvals. In collaboration with the TGA and on behalf of pharmaceutical companies, Phillip directed two major drug withdrawals. More recently, Phillip was a senior clinical trial and regulatory affairs advisor to an Australian company which developed a live virus for the treatment of melanoma.

About the Authors

Judy Wilyman

JUDY WILYMAN, PHD has expertise in the historical control of infectious diseases. She has spent many years investigating how the risk of these diseases was reduced in all developed countries by 1950, as described by the leading public health experts of the 20th century.

In her research she has also described the direct dose-response correlation between the increased use of vaccines and the decline in children's health.

In 2015, Judy completed a PhD titled *A Critical Analysis of the Australian Government's Rationale for its Vaccination Policy* that concluded that the government's claims of safety and efficacy for the childhood vaccination program were not based on hard evidence. This research included examining the scientific rigor underpinning government claims that vaccines are 'safe and effective.' These claims are based on gaps in the science, not hard evidence. That is, the empirical science that is used to prove the safety and efficacy of vaccines has never been done.

Since 2015, she has experienced censorship in the debate of

vaccination in both the mainstream media and the official channels for public debate. Her PhD is available as a download from the University of Wollongong's website and to date has had 35,800 downloads. In March 2020, just as the world was facing lockdowns and social distancing, Judy published her book titled: *Vaccination: Australia's Loss of Health Freedom.*

Now with the COVID19 pandemic scandal unfolding, Judy is in a key position to explain how and why by definition, there is no global public health emergency, and the alleged pandemic has been based on false science.

Saeed Qureshi

Dr. Saeed Qureshi has a Ph.D. in fundamental science (chemistry) specializing in analytical chemistry, which covers the science of substances' isolation, identification, characterization, purification, tests developments, validation, and their uses.

As a senior research scientist for 30 years with Health Canada, Dr. Qureshi conducted experimental studies relating to drug applications for product marketing—undertaking hands-on experimental (scientific) studies for both in vitro and in vivo (animal/human) evaluations.

He has extensively published in peer-reviewed journals and made numerous invited national and international presentations on the subjects.

He is an accomplished scientist from a regulatory organization, as reflected by several high-profile awards he has received, such as (1) the Lifetime Achievement Award (2015, Indus Foundation, India); (2) the 2007 Deputy Minister's (Health Canada) Award of Excellence in Science; (3) Excellence in Science Award (2007, Health Canada).

His expertise, experience, and work may be followed from his blog: https://bioanalyticx.com/

John O'Sullivan

JOHN IS CEO and co-founder (with Joe Olson and Dr Tim Ball) of Principia Scientific International (PSI). John is a seasoned science writer and legal analyst who assisted Dr Ball in defeating world leading climate expert, Michael 'hockey stick' Mann in the 'science trial of the century.'

O'Sullivan is credited as the visionary who formed the original 'Slayers' group of scientists in 2010 who then collaborated in creating the world's first full-volume debunk of the greenhouse gas theory.

He currently co-hosts the Sky Dragon Slayers weekly show on TNT Radio.

Robert Beatty

ROBERT HAS OVER twenty years broad experience in mine operations and planning for open cut and underground, coal and metalliferous mines. Operational experience was gained between 1965 and 1980 at middle and general management levels within the Australian Mining Industry at; Broken Hill, Tennant Creek, Blackwater, Saraji, South Blackwater, Lithgow and Atherton.

He was external Director of Mining with Sinclair Knight & Partners from 1981 to 1989. Since 1980, he operated as Principal of R.A. Beatty and Associates Pty Limited mine consulting engineers. The firm provided project management, mine planning, and research & development services under the BOSMIN® trade mark. In 1992, he was instrumental in successfully promoting the concept of ICs INTERNATIONAL which harnesses the combined efforts of several individual resource based consultants. He is currently a principal of that organisation.

In 1994, the University of Queensland recognised his lateral thinking abilities by appointing him an Honorary Research Consultant in recognition of his work on terrestrial evolution, planet orogeny and climate influences, and he has a personal interest in cosmology, investing, and gardening. He has recently written several papers which are published on the Principia Scientific International (PSI) website.

Preface

It is dangerous to be right in matters on which the established authorities are wrong.
—Voltaire

MEDICINES ARE EVERYDAY items and have become part of daily life use. It should not be like this, but unfortunately, it is so. It has almost become a high-priced version of the food or nutrients business for maintaining a "normal" or "healthy" life. They are effectively promoted as healthcare products.

The recent episode of the COVID-19 pandemic shook people's trust in the medical profession and industry, which were once considered some of the most trusted and respected professions and businesses. Several concerns created this mistrust.

1. The existence of the COVID-19 virus, its illness/infection, and spread or pandemic without providing standard and expected scientific support.
2. Lack of alternate or second opinion or option to evaluate the diagnosis critically.
3. Hugely exaggerated projections about the severity, spread, and number of deaths worldwide with arbitrary assumptions and computer modeling.
4. Developed treatments, vaccines in particular, with rushed and inept clinical trials and their protocols and endpoints.
5. Draconian measures in imposing protective remedies to otherwise healthy people.
6. Promotion of "science" by the Main Stream Media (MSM) who censored any alternate view, even from world-recognized science leaders.

The question is, why? Science had high regard, so why the current mistrust? Recently, even renowned scientists describe the published material in reputed scientific journals as uunreliable and nonreproducible.

Much ink has been spilled over the "replication crisis" in the last decade and a half, including here at Vox. Researchers have discovered, over and over, that lots of findings in fields like psychology, sociology, medicine, and economics don't hold up when other researchers try to replicate them.[1]

In June 2020, in the biggest research scandal of the pandemic so far, two of the most important medical journals each retracted a high-profile study of COVID-19 patients. Thousands of news articles, tweets, and scholarly commentaries highlighted the scandal, yet many researchers apparently failed to notice. In an examination of the most recent 200 academic articles published in 2020 that cite those papers, Science found that more than half—including many in leading journals—used the disgraced papers to support scientific findings and failed to note the retractions.[2]

Forecasting for COVID-19 has failed.

Epidemic forecasting has a dubious track-record, and its failures became more prominent with COVID-19. Poor data input, wrong modeling assumptions, high sensitivity of estimates, lack of incorporation of epidemiological features, poor past evidence on effects of available interventions, lack of transparency, errors, lack of determinacy, consideration of only one or a few dimensions of the problem at hand, lack of expertise in crucial disciplines, groupthink and bandwagon effects, and selective reporting are some of the causes of these failures.[3]

---

[1] https://www.vox.com/future-perfect/21504366/science-replication-crisis-peer-review-statistics

[2] https://www.science.org/content/article/many-scientists-citing-two-scandalous-covid-19-papers-ignore-their-retractions

[3] https://www.ncbi.nlm.nih.gov/pmc/articles/PMC7447267/

> *Why the Doomsday COVID models were so badly wrong?*[4]

The recent introduction of mandatory vaccination worldwide for young and old without appropriate safety and efficacy profiles has raised serious concerns about medication development and administration practices.

In addition, it led to mistrust in authorities for approving such products.

> *Out of nearly 2,000 U.S. nurses surveyed on Medscape (WebMD's sister site for health care professionals) between May 25 and June 3, 77% said their trust in the CDC has decreased since the start of the pandemic, and 51% said their trust in the FDA has decreased. Similarly, out of nearly 450 U.S. doctors surveyed in the same time period, 77% said their trust in the CDC has decreased and 48% said their trust in the FDA has decreased.*[5]

The traditional view is that professionals, particularly physicians and related medical experts, are qualified to know and understand the complexities of medicines. This is because they often have worked for decades with medications.

This lack of trust in authorities requires a critical evaluation of regulatory practices and the employed underlying scientific concepts and practices. Accordingly, this preface explores the potential causes of the public and the medical community's concerns by taking a nontraditional approach.

Perhaps the biggest tragedy of our time is that, although working with chemical science or chemistry at its core, experts present it as the new, modern, and advanced science of medicine. This view is presented by calling and labeling potent chemicals, often

---

[4] https://www.telegraph.co.uk/global-health/science-and-disease/doomsday-covid-models-badly-wrong/

[5] https://www.webmd.com/lung/news/20210609/trust-in-cdc-fda-took-a-beating-during-pandemic

hazardous and poisonous, as medicines. Rather than following well-established (chemical) principles, they are studied under different names such as medicine, pharmacology, toxicology, pharmaceutics, genetics, molecular biology, virology, etc.

This book, like those related publications identified herein, helps demonstrate just how far down the rabbit hole we have all had to travel during the COVID-19 pandemic in search of understanding the mass insanity.

Critical reasoning skills have been at a premium. *We The People* were betrayed by those we trusted most. From our doctors, elected representatives, the mainstream media—there is little doubt that they are a key part of the problem. Even our families and close friends seem to have shunned we who dared to show skepticism and do more than keep in lockstep with nonsensical and unscientific policies which have taken our society to the brink of collapse.

Almost all politicians, the media, large corporations, and the billionaire class are intent in an insane collaboration to impose mandatory injections contrary to the Nuremberg Code and while blinkered from those other therapies proven to be as good as, if not safer and more effective than, what Big Pharma is pushing.

From reading this book you will understand why we denounce the 'experts' who repeatedly lied without shame. Many thousands of lives have been taken or ruined on the pretext of a global threat from a "deadly new virus." The virus is no more harmful to the average person than a common flu bug.

For the better part of two years the global economy was decimated (with small to medium size businesses hit the worst), citizens' mental and physical health ruined, and draconian new emergency laws are robbing us of our freedoms. And what was the real purpose of this stupidity?

> *'Emergencies' have always been the pretext on which the safeguards of individual liberty have been eroded—and once they are suspended it is not difficult for anyone who has assumed such emergency powers to see to it that the emergency persists.*
>
> —F. A. Hayek (1974 Nobel Prize in Economic Sciences)

Risk Versus Benefits

We are being coerced to abrogate our bodily autonomy to the state.

This is biological communism.

Where are the experts explaining the realities of the risks versus the benefits? The more COVID-19 jabs you have, the more dangerous your situation.

The elephant in the room, since even before the UK's first lockdown, emerged on March 19th, 2020. As far back as then, the public health bodies in the UK and the Advisory Committee on Dangerous Pathogens decided that the new disease should no longer be classified as a 'high consequence infectious disease' and coronavirus was downgraded to flu level. Only AFTER that conclusion by the government's own experts was the economy shuttered and our lives subjected to perennial fear and lockdown lunacy.

The proof that the virus was officially regarded as no more deadly than a normal flu—BEFORE the tyranny unleashed—is spelt out in black and white at the UK government's website.[6]

But almost two years later and with no end in sight, the powers that be are insistent that the entire population must be vaccinated with experimental potions that fail the basic requirement of any traditional vaccine, namely, to protect the population from infection and transmission.

With these untried chemical concoctions, how many of your brain cells will die is something only time will tell. And children, of course, will be more vulnerable because they are likely to live longer.

Some experts, advisors and regulators will tell you that the risks are small.

But how can they know that?

And what is small?

They told us that the blood clotting problems were small.

As Dr Vernon Coleman, for decades Britain's more popular medical author on vaccine safety will tell you, in a normal

---

[6] https://www.gov.uk/guidance/high-consequence-infectious-diseases-hcid

experiment with a new drug, doctors should look and check for all the possible problems before releasing a drug for widespread use.

But the COVID-19 jabs were rolled out to billions without anyone having the faintest idea what will happen. The people were the test subjects.

Only if we arm ourselves with knowledge, do we stand a chance to make sure this never happens again.

Publisher's Note

Its weakness has always been that those who choose the lesser evil forget quickly that they chose evil.
—Hannah Arendt

YOU, DEAR READER, like the rest of us mortal souls, are in an unenviable situation. You must make decisions affecting your future—and the futures of those under your care—in the presence of incomplete and corrupt information.

We like to think of ourselves as rational creatures, but the simple fact is: we are not. We are all subject to wrong-thought, bias, prejudice, cognitive dissonance and corrupt influences. But daily, we need to act. The best we can do is take our best shot in navigating disinformation and misinformation—and accept the consequences.

Imagine an alternate reality where the COVID-19 plandemic was bad business for large companies like Yahoo, Google (Alphabet), Facebook (Meta), Instagram, Zoom, Microsoft, Twitter, Cisco (WebEx), Amazon, WalMart, Costco, TikTok (ByteDance抖音), Pfizer, Johnson & Johnson (Janssen), CVS Health, Novartis, Merck, Astra-Zenica, Eli Lilly, BiNTech and many, many others.

Would the drumbeat of panic have been heard around the world to such a degree? These companies sped up radical changes in the world—to their benefit. The transfer of wealth was incredible and irreversible.

This book represents our attempt to present information and a point-of-view based on logic, experience, research and honest attempts to minimize nonsense. We have a track record of success, but cannot make decisions on your behalf and if we could, we wouldn't. We'll leave totalitarianism to the bureaucrats and dictators.

Some of your most basic decisions include who to listen to and who to trust. Please don't take our word for anything. Think about what we've discovered and look into things yourself. And if we're wrong, as always, we want to hear about it.

That said, this book was written before the mainstream media could no longer ignore huge numbers of vaccine deaths and injuries and before Elon Musk, Matt Taibbi, Bari Weiss, Alex Berenson (who are far from being libertarian or conservative) and others unveiled the corruption of Twitter—which, as far as I can tell—100% confirm our conclusions. And, what happened at Twitter was surely duplicated at the other social media giants.

While speaking of unveiled corruption, in my opinion, Dr. Anthony Fauci is a mass murderer along the lines of Pol Pot, Joseph Stalin and Chairman Mao Zedong.

I draw this conclusion not only based on his support of the dangerous genetic therapies he calls vaccines and the illegal (and despicable) support of Gain of Function (GoF) research (which, in the case of coronviruses could accurately be called Gain of Murder because the 'function' optimized is transmission and lethality)—his corruption goes back to the phony HIV/AIDS crisis that started in the San Francisco bathhouses in the 1980s.

This weasel, dissatisfied with being the highest-paid US government employee[7] with an extravagant pension,[8] gets patent and invention royalty payments from recipients of government grants he influences—including Moderna. How much? Enough to achieve an estimated net worth of USD $12.6 million. This includes an increase

---

[7] Fauci's yearly salary in 2022 was USD $480,654.

[8] Fauci's pension will start at $333,745 yearly and will increase with cost-of-living adjustments.

of about $5 million during the COVID years 2019-2021.[9] Not bad for a so-called public servant.

> *Last year, the National Institutes of Health—Anthony Fauci's employer—doled out $30 billion in government grants to roughly 56,000 recipients. That largess of taxpayer money buys a lot of favor and clout within the scientific, research, and healthcare industries.*
>
> *However, in our breaking investigation, we found hundreds of millions of dollars in payments also flow the other way. These are royalty payments from third-party payers (think pharmaceutical companies) back to the NIH* [US National Institute of Health] *and individual NIH scientists.*
>
> *We estimate that between fiscal years 2010 and 2020, more than $350 million in royalties were paid by third-parties to the agency and NIH scientists—who are credited as co-inventor.*[10]

Does Fauci's net worth include the estimated $2 million of his wife, Dr. Christine Grady?

Who knows?

Incidentally, at the time of this writing, Christine is Chief of NIH Bioethics—the supposed conscience of public health.

> *The Department of Bioethics is a center for research, training, and service related to bioethical issues. The Department conducts conceptual, empirical, and policy-related research into bioethical issues; offers comprehensive training to future bioethicists and educational programs for biomedical researchers and clinical providers; and provides high quality ethics*

---

[9] https://openthebooks.substack.com/p/breaking-faucis-net-worth-soared

[10] https://www.openthebooks.com/substack-investigation-faucis-royalties-and-the-350-million-royalty-payment-stream-hidden-by-nih/

consultation services to clinicians, patients, and families of the NIH's Clinical Center and advice to the NIH IRBs, investigators, and others on the ethical conduct of research.[11]

How cozy. Also, it's worth noting that Dr. Grady's doctorate comes from holding a Ph.D. in philosophy from Georgetown University. Maybe she's the highest-paid philosopher in the world.

Well-played, Dr. Grady.

One thing a decision-maker needs to cultivate is a sensitive ability to detect liars. There are many vested interests, both blatant and hidden.

For me, it works well to automatically distrust politicians, bureaucrats, anyone related to big-Pharma, billionaire elitists, globalists, academics, movie stars, welfare state freeloaders, race-baiters, world-trotting noble-cause-corrupt climate activists and the mainstream media. If they're accidentally right about something, to be believed, they must provide extraordinary evidence.

Otherwise, they can peddle their toxic spew elsewhere.

While older people with complicating factors (like obesity) are vulnerable to illness and *might* benefit from a proven treatment, there is no logic to injecting healthy youngsters with experimental mRNA concoctions. This is criminal.

In military boot camp, I remember the assembly line of injections, nine vaccines in all, which made me very sick. But, that was part of the point.[12] Sick people are easy to manipulate and brainwash. And, it weeds out the weak—it's an example of getting stronger after surviving something deadly. Thank you, USAF, for trying to kill me, but failing.

One of the problems of our modern, big-data era is that regardless of what you want to believe—no matter how illogical or insane—you can find supporting sources on the web. We need to be

---

[11] https://bioethics.nih.gov/about/index.shtml

[12] https://www.health.mil/MHSHome/Military%20Health%20Topics/Health%20Readiness/Immunization%20Healthcare/Vaccine%20Recommendations/Vaccine%20Recommendations%20by%20AOR#NORTHCOM

suspicious of everything, including ourselves. Especially ourselves.

Give yourself permission to measure alternate facts and opinions and change your mind.

Recently, I became intrigued by the idea that viruses don't exist. The idea seems absurd at first, but when you fully understand the intellectual corruption of progressive activists profiting from the human-caused climate change hoax, it makes one wonder about other 'established' sciences. Anything is possible. There are many knowledgeable doctors and researchers who make the no-virus claim and they are worthy of your attention—even if you decide to disagree with them.[13]

Going further beyond whether viruses exist or not, I think there is a rational strategy for health. Your body is constantly under attack by many different things—like toxins in the air, the water and your food. What else? An area that begs for study is the effects of terrestrial, solar and cosmic radiation, including the upcoming 5G radiation.

Regardless of the threat, the best thing you can do is take care of your *terrain*. Your body. And, because the body is so influenced by your mind, you should take care of your thinking, too. The placebo effect is real, but so is the nocebo affect. Your mind and what you think can cause you to be unhealthy. This fact is exploited by many global forces. Fear is powerful. Fear kills.

Do your best to breathe clean air, drink clean water, and eat clean foods—and think clean, positive thoughts. The best strategy is to make sure your immune system is fully tuned up and ready for action.

Take care of your terrain and the miracle of the human body will serve you best. You won't live forever, but you will live your best life.

All humans, including you and me, are deeply flawed creatures. The only way we can achieve our highest calling is via liberty. Without the freedom to be wrong in the eyes of others, we will not

---

[13] Search for Drs. Mark and Samantha Bailey and see what they have to say. Read *Virus Mania*. From there, take a look at the others, like Dr. Tom Cowan, Stefan Lanka and many others. As always, weigh the information and decide for yourself.

have the freedom to be right.

We at Principia Scientific International and Stairway Press want what is best for you, your families and communities and for you and yours to live a life of liberty, peace and prosperity.

We are helping as much as we can.

Please pay it forward.

—Ken Coffman
Publisher, Stairway Press

US Center for Disease Control (CDC) Moving the Goalposts

Vaccination (pre-2015): Injection of a killed or weakened infectious organism in order to prevent the disease.

Vaccination (2015-2021): The act of introducing a vaccine into the body to produce immunity to a specific disease.

Vaccination (Sept 2021): The act of introducing a vaccine into the body to produce protection from a specific disease.

Table of Contents

Chapter 1—The Model of the Germ Theory Successful in Removing the Risk from Infectious Diseases

WHEN OUTBREAKS OF infectious diseases are observed in diverse communities it is noted that multiple causal factors are responsible for the pathogenesis of disease in individuals. In other words, infectious agents are necessary for disease to occur but exposure to them does not always result in disease; exposure to these agents can result in a diversity of health outcomes.

Therefore, the assessment of the severity of each infectious agent in the community cannot be made outside the ecological context. There are a range of health outcomes that can arise after exposure (infection) to the microorganism, and the health outcome is dependent upon the host, environment, and agent characteristics. In other words, the ecological context.

This concept is illustrated in the Epidemiological Triangle (see Figure 1) that was used to control infectious diseases in developed countries by 1950.

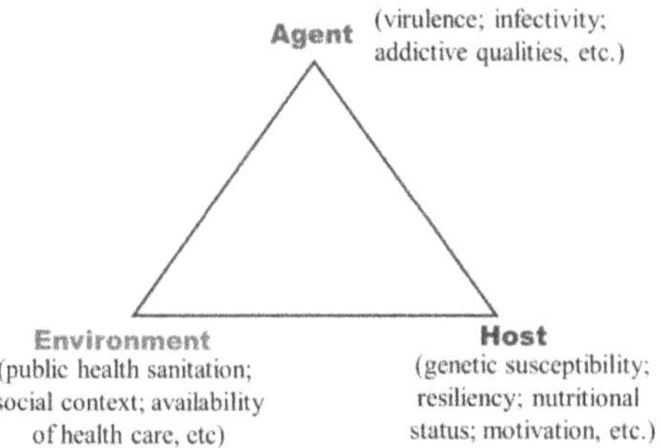

Figure 1: The Epidemiological Triad of Agent, Host and Environmental Factors[1]

In addition, humans adapt to their environment through a process known as homeostasis. This is not a characteristic that humans are born with. Homeostasis develops through interactions with the environment, and it is how the immune system builds up resistance to different diseases. The immune system interacts with the environment and other body systems to develop protection against disease.

To demonstrate the success of this model of the germ theory in removing the risk from infectious agents in developed countries, I (Judy Wilyman) have used Australia as the case study.

The Removal of Risk due to Infectious Diseases in Australia

The infant mortality rate in Australia had declined significantly by 1950. This was primarily due to improvements in sanitation, hygiene, nutrition, breast feeding, smaller family sizes, less crowding and improved infrastructure. This reduction in risk of death and illness due to infectious diseases by 1950 occurred *before* mass vaccination programs were introduced. In 1950, only two vaccines were in voluntary use in vaccination programs in Australia, diphtheria and smallpox, and these diseases did not decline any more quickly than other infectious diseases.

Measles, whooping cough (pertussis) and influenza were removed from the National Notifiable Disease list in Australia by 1950. That is, doctors were no longer required to notify the government of cases of these diseases because most people did not die or get hospitalised from these diseases.

It can be concluded that any influence of a vaccine in the decline in the infant mortality rate was secondary to the influence of changes to the environment and lifestyle that were brought about with social (ecological) medicine. That is, economic and political decisions made by the government to improve the environment, lifestyle, and quality of healthcare. Whilst outbreaks of infectious diseases still occurred after 1950, the risk of death and severe illness was low to negligible to the majority of Australian children, and long-term herd immunity was gained by natural infection during childhood, with mild or asymptomatic infections in most children.

New vaccines were introduced in Australia in vaccination programs from 1952-1990 but participation was voluntary and without coercive government strategies. Infectious diseases continued to decline as living standards and education improved. Despite the low infant mortality rate in the early 1990s and the lack of significant risk from infectious diseases in Australia at this time, the government re-introduced the national notification system to report cases of some infectious diseases. At the same time, the government implemented a new coercive strategy to increase the vaccination rates in the population and to expand the recommended schedule of vaccines.

The government re-introduced the notification of 'cases' of childhood infectious diseases, even though it was established in 1950 that cases were not an indication of the risk of these diseases to the community. This policy change was in response to a WHO directive to all member countries and *not a recommendation by the Australian government* based on characteristics that were specific the Australian ecological context. Notification of cases of disease in 1990 enabled the media to use this information to frighten the public about the risk of these diseases so that more parents would vaccinate their children.

The goal set by the WHO was to increase childhood vaccination rates to ninety percent even though there was no significant threat from infectious diseases in developed countries at this time. This means that vaccines were promoted to the public as health policies based on increasing the vaccination rates in the population, not on the health benefits that were being observed in the population. This criterion also ignored the risk from the vaccine and combination of vaccines that were being recommended to infants.

An increase in vaccination rates in the population is not the same as improving the health of children which is the promoted outcome of vaccination programs. Governments were promoting the expanded vaccination program in 1990 with financial entitlements to doctors and parents based on the assumption that high vaccination rates lead to improved health in children. After 30 years of this expanded vaccination program, children's health has significantly declined with this 'health' policy. Close to 50% of children in all developed countries now have a chronic illness or permanent disability.[1]

These diseases have increased in a direct dose-response relationship to the increased use of vaccines. This is significant because it is a strong indication of a causal link between vaccines and these diseases, and governments have never done the properly designed studies that would prove or disprove this causal link. This is called undone science in vaccination policies.

This decision to expand the vaccination program was a universal directive to all WHO member countries, (both the developed and developing countries), and it did not provide a risk/benefit assessment for the use of each vaccine in different countries. That is, in different environments and host characteristics. In addition, the benefits of vaccines to the health of the Australian community have been evaluated up to 1995 by observing the decline in the infant mortality rate. This surrogate has been used even though it is known that mortality (death) rates are unable to inform authorities about the morbidity (illnesses) associated with introducing each vaccine.

The well-being and quality of life of individuals in communities cannot be accurately measured by using mortality rates alone. After 1995, governments used the surrogate of increased vaccination rates in the population. But again, this surrogate does not inform of improved health in the community,

---

[1] https://childrenshealthdefense.org/news/chronically-ill-children-who-is-sounding-the-alarm/

only of the percentage of people vaccinated. Government's use the words 'immunisation rates' and 'vaccine-preventable diseases' to convince people to use the vaccines. These words are being misused to imply immunity is gained after every vaccination without any harm and to imply that vaccines do prevent infectious diseases. Neither of these implications are supported with empirical evidence. That is, direct observations of their effects in individuals and in community protection.

Empirical evidence is extremely important for the foundation of mandatory medical interventions in healthy infants and adults. There is now overwhelming empirical evidence of the increase in chronic illnesses and disability in Australian children (and in other countries with high vaccination rates) that has occurred at the same time as governments introduced coercive vaccination campaigns. These campaigns raised the participation rates, as well as the number of vaccines, that were being recommended in government vaccination programs.

A government that uses infant mortality rates and vaccination rates alone to inform authorities of the health of the community, will never detect or recognize a significant correlation between vaccines and increased chronic illness in the population. If you are not looking for a pattern you will not find it, and this is the false science that has underpinned government vaccination programs for decades.

The Risk of Influenza in Australia

Mortality data for influenza indicates that the number of deaths for children under 5 years of age in Western Australia is zero-to-three deaths per year. [4] This has been the number of deaths for the last 4 decades up to 2008, and it is similar to all other Australian states. [4] National deaths from influenza in children under 5 have been between zero and three annually since 1977. [4] These statistics do not justify the general vaccination of children for influenza. [5]

It is stated in the Cochrane review of influenza vaccines that the consequences of influenza in children and adults is mainly absenteeism from school and work. [6] The hospitalization and mortality data shows that the risk of complications and deaths from influenza is greatest in people over 65 years old and that there is an increased risk of complications from influenza in children under 2. [4]

The assessment of the risk of this disease should also include the morbidity from complications of influenza and an assessment of the social circumstances surrounding hospitalized cases.

Social conditions should be assessed with cases of this disease because the incidence of infectious diseases increases with poor living standards and other social factors such as nutrition. [7]

Currently this data is not reported. [1d] It should also be noted that a decision to vaccinate all children for influenza should not be based upon data from other countries as local factors such as living conditions, nutrition, available healthcare, and patterns of childcare will affect the benefits of using the vaccine. [5] Yet the Western Australian Health Department has based its promotion of childhood influenza vaccination on data from other countries. [1b]

It is observed that the attack rates for influenza are consistently high in children during annual outbreaks. [5] However, even when the attack rates are 20-30% it is known that the majority of these children make a full recovery and discomfort is the main symptom of illness. [5]

A recent survey of US pediatricians illustrated that 43% actively opposed the universal vaccination of children and 27% were unsure. [5] In addition, 50% of pediatricians were concerned about the safety of the inactivated vaccine. [5]

Complications of influenza include acute otitis media, croup, bronchitis, pneumonia, and other respiratory diseases such as asthma—and most children with these conditions are not hospitalized. It is observed that infants under 3 years and young children with underlying medical problems are at highest risk of being hospitalized. [5] The highest risk of influenza-associated hospitalization is in infants under 6 months of age, yet the inactivated influenza vaccine is only licensed for use in children 6 months and over. [5]

Today's children receive multiple vaccines and inclusion of the influenza vaccine results in some children receiving up to 14 vaccines before five years of age. [1c] The combination of multiple vaccines must be considered when weighing up the risks of diseases, as vaccines contain antibiotics, preservatives and aluminium adjuvants that are known allergens and neurotoxins. [10]

The cumulative and synergistic effects of the increased number of vaccines must be considered. It is also necessary to determine how effective the vaccine is in preventing influenza in the community.

Safety and Efficacy of Influenza Vaccine

Influenza is a disease that is caused by hundreds of strains of virus; however, the vaccine only protects against one to three strains depending on the type of vaccine used. [1e]

The government uses two definitions to describe the effectiveness of the vaccine. The term efficacy is used to describe how well the vaccine protects against the 3 strains of influenza covered by the vaccine. For example, the current vaccine in 2008 protects against Type A (H1N1), Type A (H3N2) and Type B (1e). The term effectiveness of the vaccine is used to describe the ability of the vaccine to protect against Iinfluenza-Like Illness (ILI), that is, the influenza cases that are not laboratory confirmed and the strain of virus is unknown. [1e]

Therefore, some ILI will be caused by strains of virus that are not present in the vaccine and these cases will not be recorded in the surveillance of influenza. So, the only real indicator of whether vaccine programs are reducing the incidence of influenza in the community is to monitor the hospitalizations and death due to all influenza-like illness (ILI) each year—not just a percentage of cases that are sub-typed for strains covered by the vaccine. At present the WA government is reporting only on some hospitalizations that are sub-typed for the strains of influenza covered in the vaccine to support its policy. [1d]

It is possible that because there are many viruses causing influenza illness in the community reducing the circulation of 2 or 3 will not reduce ILI in the community as other strains of influenza will infect. This is another reason why it is important to analyze hospitalization and mortality data to ensure this program is achieving its outcomes. An assessment of this data will confirm whether predicting the most severe strain of influenza virus a year in advance is a successful strategy.

In Australia the flu vaccine has been offered free to people 65 years and older since 1999. This program has had an uptake rate of 79%. [4b] The strongest evidence for the effectiveness of this campaign would be an analysis of the hospitalization data and deaths in this age group since the program started ten years ago. This analysis has not been published or presented as evidence in the formulation of current influenza policies in 2008. [3a]

A recent Cochrane Review of all the studies conducted on the effectiveness of influenza vaccines in children stated that the efficacy of inactivated vaccines for children under two (against strains contained in the vaccine) was similar to placebo, that is, not effective at all. [6] It should also be noted that the Cochrane Review states that neither type of influenza vaccine—inactivated or weakened influenza viruses (nasal sprays) were good at preventing ILI in children over 2. [6]

It was also concluded in the Cochrane Review that due to the variability in study design an analysis of safety data for influenza vaccines in children was not feasible. [6]

Despite significant coverage of the influenza vaccine in the Australian community for many years, both in the elderly and in workplaces, 2007 was described as a severe flu season with notifications being 3.4 times the 5-year mean. [3c] In Western Australia it was described as being the worst flu outbreak in four years. [(12] This evidence is not an indication that influenza vaccine is reducing the incidence of this disease in the community.

Although notifications for this disease are highest in the 0-4 year age group, this is not a reflection of the severity of the disease in the population. This is because influenza is only considered a serious disease in the elderly and immune compromised and the majority of children and adults make a complete recovery after several days. [5] [13] [7] It is hospitalisations and deaths that indicate the risk of influenza, not cases of flu from which most people recover and get immunity.

The Health of Australian Children

Our knowledge of the effects of vaccines has now been collected for over 100 years. It is important to look at the ecological health of the population as well as the statistics collected over this time to ensure this procedure is safe. Statistics can hide many variables. The ecological evidence is showing that the health of children has not improved as the number of vaccines on the childhood schedule has increased. [4c]

Chronic illness in children has risen dramatically in the last two decades and this coincides with the government's push to increase vaccination rates in Australia to 95% with the addition of many more vaccines to the childhood program. This was started with the implementation of the Immunise Australia Program in 1993. [3b]

Children's health and the health of society are dependent upon scientifically proven preventative policies. It has not been proven that vaccines are not a cause of chronic illness in children. The increase in autoimmune diseases in dogs and cats has already been linked to vaccines and we must consider this same possibility in our children. [14] [15]

The Evaluation and Promotion of Influenza Vaccination in Western Australia

Australia has adopted the guidelines set by the Centre for Disease Control's Advisory Committee on Immunization Practices (ACIP,

USA) which state that annual vaccination of all children aged 6 months to 4 years should continue to be a primary focus of vaccination efforts because these children are at higher risk for influenza complications compared to older children. [16] The ACIP also recommends that annual vaccination be administered to all children aged 5-18 years. [16]

Western Australia adopted this initiative in autumn 2008. The WA Health Department promoted free childhood influenza vaccines through an advertising campaign in the media. The advertisements used the deaths of three children in 2007 to suggest influenza is a serious risk to all children. [12] Further examination of these deaths revealed that the cause of death for these three children was inconclusive and still subject to a coroner's report at the time of the media promotion. [1b]

The Director of the flu campaign stated the information on these deaths was restricted to the public, yet the information was used in a state-based media campaign. [1b] These deaths represent anecdotal evidence of the risk of influenza to children and it was revealed in the media that only one of the children was confirmed with Influenza A as opposed to all three children that the vaccine advertisement had implied. [12] All three children were confirmed to have bacterial pneumonia, but this was only provided in the fine print underneath the advertisement.

Discussion

Children are considered the main transmitters of influenza in the community. [5] Heikkinen et al (2006) therefore suggest it is logical to assume that vaccinating children would lead to substantial reductions in parental work loss due to caring for sick children. [5] They also suggest it could lead to decreased morbidity and mortality in the elderly. This is an assumption that ignores the theory of opportunistic infection. It is possible that because there are many strains of influenza viruses circulating in the community, reducing the incidence of 2 or 3 will still leave individuals susceptible to other circulating strains that will increase.

The fact that severe outbreaks of influenza are still being observed despite vaccination campaigns would appear to support this theory. In other words, matching for 2 or 3 strains of influenza virus is not affecting the incidence of influenza-associated illness (ILI) or deaths in the community because there are many other strains that cause influenza-like infection.

Heikkinen et al (2006) state that the 'average efficacy of inactivated influenza vaccine is approximately 70-80%'. [7, p.226] That is, it will reduce 70-80% of influenza caused by the strains of virus covered by the vaccine. These authors then imply that the vaccine will reduce rates of illness, influenza-associated complications, and hospitalizations among vaccinees by the same percentage. However, they admit that the overall effectiveness of the vaccine will be reduced substantially in the community because of the other respiratory viruses/bacteria (including other strains of influenza) in circulation. Therefore, predicting the percentage reduction of influenza in the community is not possible. [5] Empirical evidence is essential in determining the reduction in illness that will result in the community because of the many variables involved.

The evidence being used by policymakers regarding the efficacy of influenza vaccine is derived from random and quasi-randomised controlled trials and observational studies. [17] Jefferson (2006) explains that many of these studies are of poor methodological quality and are known to be affected by bias and confounding factors. [18] As a result, these studies provide inconclusive evidence on vaccine effectiveness which leaves the issue open to debate.

To provide more conclusive evidence on vaccine effectiveness in the community, the government must provide the hospitalization and mortality data of all influenza-like illness monitored over the period of vaccine usage. Until this data is published, the effectiveness of influenza vaccine will be unknown.

Conclusion

Influenza has been promoted to the public in Western Australia as a serious risk to children even though the influenza-associated mortality for children is described as extremely low. The children at highest risk from influenza are children under 6 months of age and the vaccine is not licensed for this age group. In addition, the inactivated flu vaccine has been described as ineffective for children under two—the age group with the highest complications to flu. Childhood vaccines have also been described as having low effectiveness against ILI. The other reason for vaccinating children is to see whether it lowers the transmission of influenza in the community.

Statistics can hide many variables, so it is important that the public is presented with information that best represents the severity of this disease in the community—not the incidence in the community because this disease is not severe in all age-groups. At present, the government only reports on a percentage of sub-types of influenza that are hospitalized. In order to show the effectiveness of the vaccines against flu in the community the government must report on all cases of ILI that are hospitalised.

Monitoring ILI will inform us whether the theory of selecting to protect against 2 or 3 strains of influenza is effective in reducing the hospitalisations and mortality of influenza in the community: the desired outcome of vaccination programs. It is possible that targeting 2 or 3 species only allows a space for one of the many other influenza viruses/bacteria to cause infections. In this case there will be no reduction in hospitalizations due to influenza-associated illness.

Other evidence that should be used in the risk analysis of influenza vaccination is the ecological evidence in the population. In the case of influenza campaigns and children's health, there are two ecological trends that are being observed:

1) communities are still experiencing severe outbreaks of

influenza despite vaccination campaigns in the elderly and in workplaces.

2) children's health has declined as the use of vaccines has increased. If scientists cannot prove vaccination is not the cause of chronic illness, then it is unethical to continue adding vaccines to the childhood schedule.

A decision to use influenza vaccine must also consider evidence regarding the effectiveness of the vaccine. The Cochrane systematic review of vaccines does not suggest this vaccine is effective in children—particularly those under two.

The evidence suggests the Western Australian Government has run a fear campaign in the media, based on anecdotal evidence to encourage parents to vaccinate their children. If the government is misrepresenting the risk of influenza to children and over-stating the benefits of the vaccine to the community, this policy could have serious consequences for children's health and society. It also undermines the independent nature and credibility of the government advice.

Vaccines are not without risk, so it is important that value judgments about the necessity for a vaccine are made from non-biased sources. The government must therefore be seen to be openly informing parents on this issue. It is also essential that governments consider the possibility that multiple vaccines in infants are doing more harm than good particularly as this link has been described in veterinary journals and it is known that individuals can be genetically pre-disposed to chronic illnesses.

This research has implications for the use of coercive and mandatory vaccination policies.

Governments need to be more selective about the vaccines they recommend on the childhood schedule and vaccination policies should remain fully voluntary unless the government can demonstrate the community is seriously at risk if a vaccine is not used.

Chapter 1 References

1) Government of Western Australia, Department of Health,

a) Media Release 15th February 2008, *Free Vaccines to Help Fight Child Influenza*

b) Communicable Diseases Control Directorate, Van Buynder P, June 2008

c) Childhood Immunisation 2009

d) WA Communicable Diseases Bulletin, Disease Watch, March 2009, Vol13, No.1

e) Communicable Diseases Control Directorate, Influenza Fact Sheet, 2009

f) WA Public Health Bill 2008

2) Friis RH and Sellers TA, 2004, *Epidemiology for Public Health Practice* (3rd Ed.), Jones and Bartlett Publishers, USA

3) Australian Government, Department of Health and Ageing,

a) National Centre for Immunisation Research and Surveillance (NCIRS)

b) Immunise Australia Program

c) Australian Influenza Report, Report No.13 Week ending 13 October 2007.

4) a) Australian Government, Australian Institute of Health and Welfare, National Mortality Database, GRIM Book Influenza, 2005.

b) Australian Government, Australian Institute of Health and Welfare, 2005, *Adult Vaccination Survey October 2004: Summary Results*, AIHW cat. No. PHE 56. Canberra: AIHW & DOHA

c) *Child Health, Development and Wellbeing: A Picture of Australia's Children* (May, 2005) www.aihw.gov.au visited 10.03.06

5) Heikkinen T, Booy R, Campins M, Finn A, Olcen P, Peltola H, Rodrigo C, Schmitt H, Schumacher F, Teo S, Weil-Olivier C, 2006, *Should Healthy Children be Vaccinated Against Influenza?,* European Journal of Pediatrics, 165: 223-228, DOI 10.1007/s00431-005-0040-9

6) Jefferson T, Rivetti A, Harnden A, Di Pietrantonj C, Demicheli V, 2008, *Vaccines for Preventing Influenza in Healthy Children*, Cochrane Database of Systematic Reviews, Issue 2, 2008, Art. No.: CD004879.

7) Burnet, M., 1952, *The Pattern of Disease in Childhood*, Australasian Annals of Medicine, Vol.1, No. 2: p. 93.

8) Eldred BE, Dean AJ, McGuire TM, Nash AL, 2006, *Vaccine*

Components and Constituents: Responding to Consumer Concerns, Medical Journal of Australia, Vol. 184 Number 4, 20th February 2006.

9) Gilbert SG, 2004, *A Small Dose of Toxicology: the Health Effects of Common Chemicals*, Boca Raton, FL, CRC Press.

10) Australian Medical Association (AMA) www.ama.com.au/FAQ visited 11.04.10.

11) Jefferson T, Rivetti D, Di Pietrantonj C, Rivetti A, Demicheli V, 2008, *Vaccines for Preventing Influenza in Healthy Adults*, Cochrane Database of Systematic Reviews 2007, Issue 2. Art. No: CD001269.

12) Western Australian Government, Dept. Health, 2008, Flu vaccination advertisement, The West Australian Newspaper, 6th July 2007. www.public.health.wa.gov.au

13) Hays JN, 2000, *The Burdens of Disease: Epidemics and Human Response in Western History*, Rutgers University Press, New Jersey/London.

14) La Rosa, W.R., 2002, *The Hayward Foundation Study on Vaccines; A Possible Etiology of Autoimmune Diseases.* www.homestead.com/vonhapsburg/haywardstudyonvaccines.html_ visited 18.01.06

15) O'Driscoll, 2006, Shock to the System; *The Facts about Animal Vaccination, Pet Food and How to Keep your Pets Healthy*, Abbeywood Publishing Ltd, 2005, Great Britain

16) US Government, Department of Health and Human Services, Centers for disease Control and Prevention, 2008, Prevention and Control of Influenza; *Recommendations of the Advisory Committee on Immunisation Practices (ACIP)*, Morbidity and Mortality Weekly Report (MMWR) 17th July 2008.

17) Rivetti A, Jefferson T, Thomas R, Rudin M, Rivetti A, Di Pietrantonj C, Demicheli V, 2008, *Vaccines for Preventing Influenza in the Elderly*, Cochrane Database of Systematic Reviews 2006, Issue 3. Art No.: CD004876

18) Jefferson T, 2006, *Author's Response to Influenza Vaccination: Policy v Evidence*, The British Medical Journal, Letters; 333:1172 (2 December), doi: 10. 1136/bmj

Chapter 2—World Wide Medical Expertise

AUTHOR SAEED QURESHI formed views working at a regulatory authority (Health Canada) for 30 years with academic training in chemistry with a Ph.D. degree specializing in analytical chemistry (i.e., the science of isolating substances and developing, validating, and applying test methods).

Working as a bench scientist and successfully applying the expertise in the disciplines above showed that the so-called science of medicine/pharmaceuticals is a fraudulent and deceptive chemistry variant.

For example, consider the current PCR test, which cannot be regarded as relevant and valid without its validation against the virus by any basic understanding of science. The test has not been validated because the virus has never been isolated, but false claims were made. The lack of these things, i.e., validated tests and the absence of the virus, make the whole story of pandemic and vaccination false and fraudulent.

Perhaps the most surprising and disappointing aspect is that the treatment had been developed on the fly fully knowingly that neither the patient population nor the testing methodology for the disease is available, with claims such as:

> *...it was a reflection of the extraordinary scientific advances in these types of vaccines which allowed us to do things in months that actually took years before.*[2]

Sad!

So the question is, what went wrong with the science or its

[2] https://www.livescience.com/fauci-vaccines-not-rushed.html

claims? Consider a common quote associated with Albert Einstein:

> *No problem can be solved from the same consciousness that created it.*

Therefore, it is time to look from a different and fresher perspective. The following discussion is provided precisely with this objective—explaining the situation and suggesting relatively simple, logical solutions considering well-researched and established fundamental scientific principles and practices.

This book is for the general public and professionals, including the healthcare industry. However, technical details are kept to a minimum but explained in simple or laymen's language. The users or patients of medicines could clearly understand the actual science and its misuse for defining illnesses and in developing and using the medication. Most of the medical literature is directed towards the select group, medical and pharmaceutical professionals, and related disciplines, based on peer-reviewed assessment, not necessarily as implied unbiased and third-party review system.

A particular shortcoming of the current peer-reviewed-based system is that it ignores the input from the end-user (i.e., the patient), other health-related science disciplines and any other third-party alternative healthcare knowledge and expertise. Thus, everything is within a select group of like-minded people, including regulatory authorities, arguably indicating self-interest and gains.

Furthermore, it is assumed and emphasized that the practice of medicines is highly sophisticated and complex. It requires multiple years of study and experience, following only a set path in a hospital environment.

This mindset, unfortunately, created a cult-like system that everyone must accept and follow at its face value. There is no room for critically evaluating or considering input from others. Instead, the select group, patronized by regulatory or bureaucratic authorities, must approve any smallest suggestion or input, otherwise be rejected, claiming a lack of appropriate scientific understanding or outside the domain of modern medical understanding.

A third-party review is the only practical way to appropriately judge professional claims and outcomes' usefulness and effectiveness and address the current unfortunate situation. In reality, the professionals should explain the science in simple language so that the patient and the third-party professionals can follow it easily. Therefore, it is incorrect to suggest that the outside reviewers must first be competent in reading the medical literature with a similar understanding, level of education and training as the professionals.

Critical review or evaluation by external experts is an expected and standard practice practically in every profession except in the medical and pharmaceutical professions. Anyone familiar with third-party reviewers of big or small appliances or service providers would attest to it. For example, the reviewers do not have to have an advanced degree or detailed understanding of a car's engineering and mechanical aspects to review the vehicle's performance. Instead, they review the car's performance against the claims made by the manufacturers. There is a critical lack of this understanding and practice in the medical and pharmaceutical professions.

For example, if one would ask medical or pharmaceutical experts how to describe the quality of a pharmaceutical product such as Tylenol or Lipitor tablets. Unfortunately, there would not be an answer as no measurable criteria are available to define medicines' quality. The only most common response is that the worldwide authorities, including FDA, have approved the products.

The working definition of monitoring pharmaceutical products' quality could be, as noted by Janet Woodcock, acting Commissioner of the U.S. Food and Drug Administration:

> *The literature does not offer a consensus definition of quality of any product or service.*
>
> *Although no unified FDA definition of drug*

> *quality for regulatory purposes has been articulated, an operational definition can be discerned from an understanding of FDA practices.*[3]

So, all medicinal products available on the market are assumed to be of quality because experts and, by extension, authorities say so, and the public has to accept them. No science-based standards or criteria are available in this respect but opinions.

A recent example is that of the virus pandemic. No one has seen or isolated a virus (directly or indirectly) in any laboratory. However, it has been declared that the virus exists and caused the illness and pandemic—strange! Moreover, governments across the globe enforced a mandate using the face mask, requiring social distances and lockdowns to protect the public from the virus. Yet, no experimental study has shown that such a measure provide any benefit. So, what is the basis of the mandates—it is an unchecked or unaudited authorities' view and experts' opinion, nothing more!

On the other hand, if the subject is studied considering relevant scientific principles and theories and evaluated logically, the medical and pharmaceutical profession will work like any other effective and useful profession. This book intends to cover this aspect by explaining to the public and patients, particularly in the most straightforward possible language, what medicine and therapeutics mean, how they work, and the basis of their development.

A casual reader may immediately be skeptical about how a non-medical and pharmaceutical professional can explain the relevant science. However, this is precisely the purpose: to provide a new and scientifically valid view that, unfortunately, medical and pharmaceutical professionals miss during their academic learning

---

[3] *The Concept of Pharmaceutical Quality*, American Pharmaceutical Review, November 2004 (https://www.drug-dissolution-testing.com/blog/files/Woodcock2004apr.pdf)

and training and, by extension, during practice.

Most of the provided discussion is based on well-established scientific principles that have been taught over decades. However, the focus will remain that content should easily be understood by medicine users and other learned medical and pharmaceutical authorities.

The terms medicines, drugs, pharmaceuticals and therapeutics are interchangeable and are used as such. However, they all are chemical compounds, most often very potent and possibly extremely poisonous. Therefore, one must understand that one is dealing with chemicals when dealing with pharmaceuticals. However, the body at the fundamental (cellular) level works as a sophisticated but exact and efficient chemical factory. Thousands of chemical reactions occur every minute in the body to break down and reassemble chemical compounds, all based on well-established and understood reactions.

Medicines, like other chemicals, become part of this massive and remarkable chemical factory called the human body. Therefore, it is essential to understand the chemistry aspects of chemicals and chemical reactions to understand medicines. However, modern-day medicine development practices lacking such knowledge amount to ignorance and shooting in the dark, resulting in problems like the current drug adverse reactions, imaginary viruses, and pandemics.

Chapter 3—Observations and Claims without Evidence

THIS BOOK PRESENTS a wealth of scientific data that shows there can be no sensible public health reason for continuing the madness. Even by February 2021, a survey of scientists[4] had admitted that the novel coronavirus is becoming endemic and will be something we live with, just like the common cold.[5]

The wisdom and reasonableness of the Great Barrington Declaration was mostly ignored. By the end of 2021 we realized the year turned out to be worse than 2020. By mid-June, around 180 million Americans had received at least one dose of the (Emergency Use Authorised, EUA) 'vaccines' and even states with the most authoritarian approaches, such as New York, had lifted most of their restrictions. Deaths went down in the Spring, but by July infections started to spike again with the new variant, dubbed Delta, less deadly, but more infectious. The government response was to tell everyone to get another shot—a booster.

The virus also showed signs of seasonality. States such as New York mandated the shot for government workers, private school staff and most indoor establishment workers. In many public venues patrons now must show proof of *full* vaccination. Outgoing NY Mayor Bill de Blasio also added a mandate for all private employers and the state mandated the jabs for all health care workers.

Mandatory vaccination thus went from being a conspiracy theory to official policy and the death rate among children and under 40's began to sky rocket with the double-jabbed appearing to be most susceptible.[6]

As Petr Svab reported in the Epoch Times:

---

[4] https://www.nature.com/articles/d41586-021-00396-2

[5] Ibid.

[6] *Joe Biden: Covid Vaccination in US will not be Mandatory* https://www.bbc.com/news/world-us-canada-55193939

> *While in 2020, about 0.05 percent of all COVID-19 deaths were children, the number rose to 0.1 percent in 2021, according to CDC,*[7] *making for a total of 678 deaths associated with COVID-19. For comparison, the agency recorded* [8] *358 pediatric deaths during the 2009-2010 flu pandemic.*[9]

Unless you view the world through the eyes of a eugenicist, nothing of any benefit has been achieved. We watched as the elite—the mega-rich, corporations, the legacy media and sold-out academia—had their wealth and status rise exponentially. Students of history know full well that the Spanish Flu pandemic was over in 18 months without lockdowns and draconian restrictions.

Likewise, the Asian flu of 1957-58[10] was a truly deadly pandemic with a far broader reach for severe outcomes than the COVID-19 hysteria of 2020. History shows the Asian flu pandemic killed between 1 and 4 million people worldwide. It was a leading contributor to a year in which the US saw 62,000 excess deaths.

By comparison to other pandemics, the COVID-19 pandemic is a drop in the ocean, as shown in the graph below.

---

[7] https://www.cdc.gov/nchs/nvss/vsrr/covid_weekly/index.htm

[8] https://www.cdc.gov/flu/about/season/flu-season-2017-2018.htm

[9] https://www.theepochtimes.com/mkt_morningbrief/2021-covid-19-recap-200-million-vaccinated-450000-dead_4190440.html?utm_source=morningbriefnoe&utm_medium=email&utm_campaign=mb-2022-01-04&mktids=2164b04e22d32ab99628001bc4a429d5&est=gzRIHlAnmS%2BMD5Zp9A%2BxWLucbX8H4PNs%2FIQAvRXs2%2FaDVjA2pODhqlVMG0ZPTzOOVshsfVnP

[10] https://www.aier.org/article/elvis-was-king-ike-was-president-and-116000-american

Figure 2. New Respiratory Disease Cases as Reported through the National Health Survey, United States, September-December 1957[a]

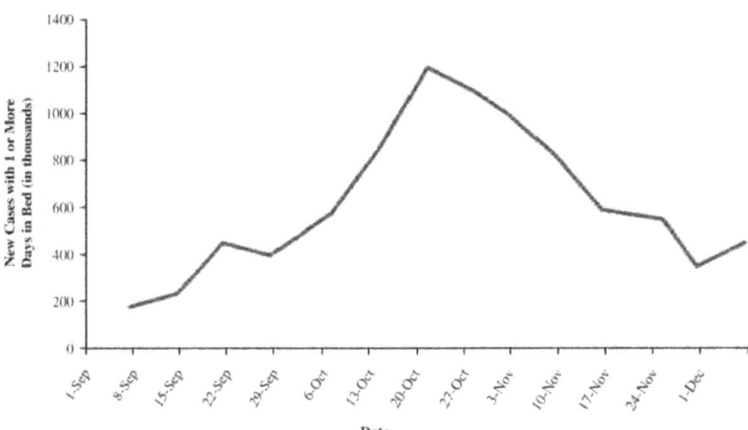

Never take anyone's word on anything until we can see with our own eyes that any theory can be of predictive value in nature, i.e. that a claim for the existence of a deadly new pathogen such as a virus can be shown (directly or indirectly) to be distinct and real under a microscope.

In the case of the claimed novel coronavirus named SARS-CoV-2, we have seen many Freedom of Information (FOI) requests for proof of a gold standard new pathogen be thwarted.

No government, no scientific laboratory, no corporate entity, NGO or otherwise has yet been able to demonstrate empirically that they can isolate this novel pathogen, perform a replication test on humans or animals and verify that an infection with the same symptoms is created.

In Chapter 4, we see a standard step by step of how the proper scientific method, handed down to us since the age of Sir Isaac Newton, is performed. It remains the standard process relied on by applied scientists in Science, Technology, Engineering and Mathematics (STEM) fields. It was abandoned in academia when post-normal science took hold.

At the center of the cycle of constant testing and evaluation is the maxim "show your work."

That means validate your conclusions by demonstrating proof that is reproducible in tests carried out independently by those skeptical of your claims.

Chapter 4—Rigorous Science

RIGOROUS DEVOTION TO the traditional scientific method is fundamental to our methodology when addressing new scientific claims.

The diagram below neatly illustrates how real science must be done. The process to follow is simple: form a hypothesis from direct observation of nature, reproduce and simulate the natural process, make predictions to test your theory. If you cannot verify your theoretical claims with hard evidence, your theory is dead and you must start over.

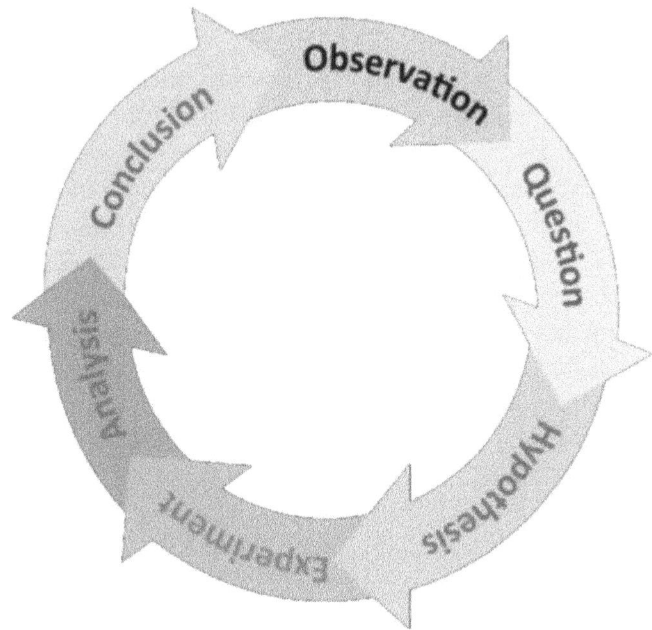

We are skeptical souls who prefer open objective and unbiased parents using accurate and balanced information on the risks and benefits. Or was it based on a fear campaign run by the media? This analysis addresses whether there is strong evidence for the claim that influenza vaccine should be recommended to all demographics and whether governments have provided evidence it is safe to use multiple vaccines in infants.

The most conclusive evidence for determining the health effects of combining multiple vaccines in infant bodies comes from long-term prospective studies on animals and humans. [2] This type of study provides evidence of the cumulative, synergistic, and latent effects of the chemicals in vaccines on children's health outcomes. A search of government and medical documents shows neither of these studies has been done. [3a]

The evidence being used by advisory committees to claim it is safe to use multiple vaccines in infants comes from short-term epidemiological (statistical) studies observing one or two vaccines at a time. [2b]

This does not give us the complete picture of how the combination of vaccines will affect children's health.

In the absence of conclusive evidence on the safety of childhood vaccination, data on the ecological health of Australian children should be used as a safety signal for this program. Ecological health is the overall change in health outcomes that are observed in the population of Australian children over 5-10 years. These outcomes should be evaluated to assess if there is a suggestive link between the use of multiple vaccines in children and a significant increase in chronic illnesses.

This evidence is provided in this analysis.

Chapter 5—Chemistry/Science for Medicines

THE BASIC UNITS of substances are elements or atoms. The most commonly referred to in biological and medical areas are Hydrogen (H), Oxygen (O), and Nitrogen (N) (they all happen to be gases at room temperature). Some examples of non-gaseous elements are Carbon (C), Phosphorous (P), Sulfur (S), Calcium (Ca), and Iron (Fe). In chemistry, they are represented with symbols (one or two first letters of their names) mainly derived from their English names, but some Latin and others.

After elements come molecules which are composed of multi-atom. For example: Oxygen (O_2), Hydrogen (H_2), and Nitrogen (N_2) molecules combining two of the same atoms. These atoms are usually not available in free form but would react with each other and form a stable molecule. Examples of molecules of multiple elements are water (H_2O), carbon dioxide (CO_2), silica, or silicon dioxide (SiO_2). Molecules having different types of atoms are called compounds.

Another way of representing molecules, aside from symbols, is by their connection or bonds, e.g., water (common name, H_2O chemical name, H-O-H structural symbol), Carbon dioxide (common name, CO_2 chemical name, structural name O=C=O), salt, or table salt (common name, NaCl chemical name, structural name, Na-Cl), glucose (common name, chemical name dextrose, formula $C_6H_{12}O_6$, structural name H-O-CH(OH)-CH(OH)-CH(OH)-CH(OH)-CH2(OH)), alcohol (common name, chemical name ethanol, formula C_2H_5-OH). Connecting lines represent bonding or bonds.+++

These are not very difficult to learn or remember because they are very systematic and logical. For example, a large group of molecules is known as hydrocarbons. Hydrocarbon means molecules of hydrogen and carbon. The simplest example of this is methane gas (home heating gas or CH_4).

```
    H
    |
H - C - H
    |
    H
```

Figure 1
Chemical Structure of One Carbon atom, Called Methane

The next is ethane gas (C_2H_6), or:

```
    H   H
    |   |
H - C - C - H
    |   |
    H   H
```

Figure 2
Chemical Structure of Two Carbon Atoms, Called Ethane

The next is propane and butane gases (BBQ gases, C_3H_8, and C_4H_{10}, respectively).

```
    H   H   H              H   H   H   H
    |   |   |              |   |   |   |
H - C - C - C - H      H - C - C - C - C - H
    |   |   |              |   |   |   |
    H   H   H              H   H   H   H
```

Figure 3
Chemical Structures of Three and Four Carbon Atoms, Propane and Butane, Respectively

The next is a variation of gas/petrol for cars which is octane, i.e., a molecule with eight carbon atoms (i.e., C_8H_{18}).

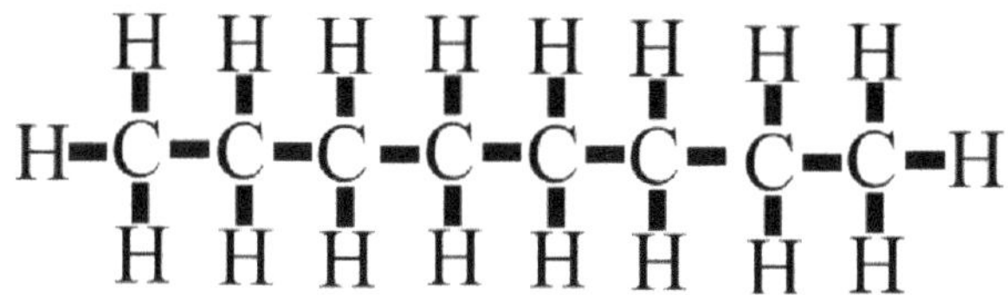

Figure 4
Chemical Structures of Eight Carbon (Hydrocarbon) Called Octane

So on and so forth.

These hydrocarbons (also commonly known as paraffin) occur abundantly in nature (e.g., crude oil). Individual members of the series have been recorded up to $C_{78}H_{158}$ in crude oil. Besides the individual names, one should note a relatively simple formula and composition describing hydrocarbon C_nH_{2n+2}, where "n" is the number of carbons in the molecules.

One can get various compounds and substances with only two atoms (carbon and hydrogen).

Adding another atom, e.g., oxygen, to the carbon and hydrogen compounds by replacing or inserting an oxygen atom into a propane molecule.

Adding a -OH (or hydroxyl) group in the base structure will result in a name as "ol" instead of "-ane." Hence, the two-carbon molecule, ethane (CH_3-CH_3), becomes ethanol CH_3-CH_2OH (i.e., our favorite alcohol).

Next, propane ($CH_3CH_2CH_3$) to propanol ($CH_3CH_2CH_2OH$), a form of washing alcohol.

If one puts the "-OH" group on each carbon on propane, i.e., $CH_2(OH)$-$CH(OH)$-$CH_2(OH)$, it is called glycerol (commonly known as glycerine).

The structure, in the vertical version, of glycerol or glycerine is as follows.

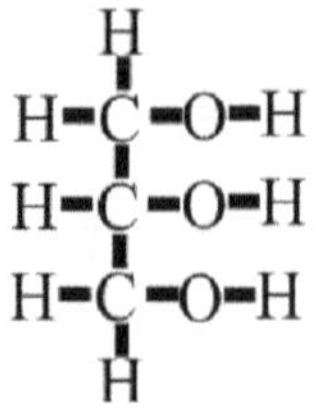

```
     H
     |
H - C - O - H
     |
H - C - O - H
     |
H - C - O - H
     |
     H
```

Figure 5
Chemical Structure of Propane with Three Hydroxyl Groups, called Glycerol or Glycerine

In addition to "–ols" there are other chemical compounds with carbon, hydrogen, and oxygen. They are called carboxylic acids or organic acids, or simply acids. The simplest example in the family is acetic acid ($C_2H_4O_2$, or CH_3COOH)—a major component of vinegar.

The structural formula is:

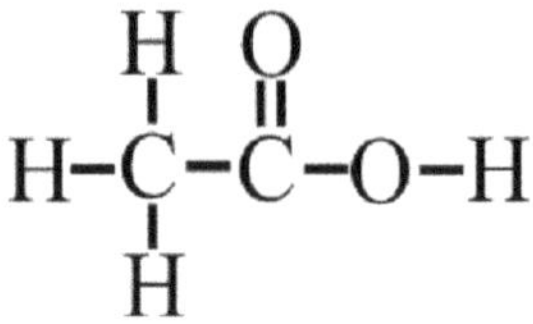

```
    H   O
    |   ||
H - C - C - O - H
    |
    H
```

Figure 6
Chemical Structure of Acetic Acid

It is built on an ethane molecule—however, the carbon atom on the right is connected to two oxygen atoms instead of three hydrogen atoms. The lines between atoms represent the bonds between the atoms, like welding joints. Every atom comes with its welding joints, also known as valencies. For example, hydrogen has one, oxygen two, and carbon four. They join according to their valencies or welding spots with different atoms. One can see that the carbon (four valencies) on the left is bound with three hydrogens (having one valency) while carbon on the right is with two oxygen atoms, i.e., four bonds. One can observe that carbon, oxygen, and hydrogen atoms fulfill the requirements of their valencies, i.e., four for carbon, two for oxygen, and one for hydrogen. These bonds dictate how atoms join and orient in space.

Keeping the "-COOH" molecule but changing the hydrocarbon component results in many organic acids. For example, let us change two ethane carbon atoms by twelve carbon atoms with the formula $C_{12}H_{24}O_2$).

```
  H H H H H H H H H H H O
  | | | | | | | | | | | ||
H-C-C-C-C-C-C-C-C-C-C-C-C-O-H
  | | | | | | | | | | |
  H H H H H H H H H H H
```

Figure 7
Chemical Structure of an Acid with a Long Hydrocarbon Chain (Lauric Acid)

This acid is called lauric acid, the major component of coconut oil. These long-change acids are commonly known as fat molecules.

Another common fatty oil or fat is oleic acid. It has an eighteen carbon backbone and is similar to lauric acid except for one notable difference: one of the carbon-carbon bonds is a double bond.

```
  H H H H H H H H     H H H H H H H O
  | | | | | | | |     | | | | | | | ||
H-C-C-C-C-C-C-C-C-C=C-C-C-C-C-C-C-C-C-O-H
  | | | | | | | | | | | | | | | | |
  H H H H H H H H H H H H H H H H H
```

Figure 8
Chemical Structure of a Long Chain Organic Acid (Oleic Acid)

A fat molecule with a double bond is known as an unsaturated one. In this case, it has fewer hydrogen atoms by 2, and would become saturated by adding two more hydrogen atoms in the molecule, it is a relatively easy chemical reaction to achieve this. The double bond is on carbon 9 and 10 (position counting starts with carboxylic acid's carbon being number 1). Margarine is usually a saturated version of unsaturated oils or fats.

Let us add another atom to the series, nitrogen (N). The nitrogen atom has three valancies; therefore, it can form three single bonds. Recalling carbon with four single bonds (methane CH_4), the similar form with nitrogen is called ammonia and is represented as NH_3. Ammonia reacts or combines with other molecules. Such molecules are called amines. For example, if one replaces one hydrogen atom on the carbon backbone of methane with nitrogen, it will be called methylamine.

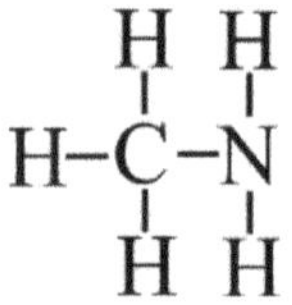

```
    H H
    | |
  H-C-N
    | |
    H H
```

Figure 9
Chemical Structure of a Simple Amine (Methylamine)

Adding carbon atoms, or hydrocarbon molecules (or chain) like carboxylic acids, forms respective amines, such as ethylamine, propylamine, etc.

Also, from a biological and medical perspective, let us add carboxylic acid as well. For example, the simplest molecule in this category is called glycine, and its formula or structure is given below.

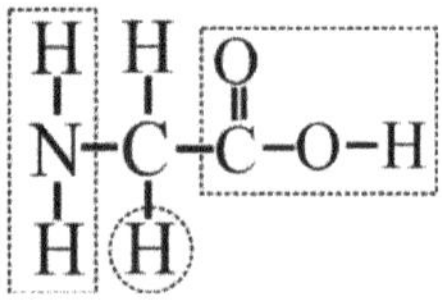

```
H H O
| | ‖
N-C-C-O-H
| |
H H
```

Figure 10
Chemical Structure of a Simple Amino Acid (Glycine)

The specific name for it, as mentioned, is glycine. However, the generic name for this class of molecules is amino acids.

Some more examples of amino acids are:

```
H  H  O          H  H  O          H  H  O
|  |  ||         |  |  ||         |  |  ||
N--C--C--O--H    N--C--C--O--H    N--C--C--O--H
|  |             |  |             |  |
H [CH3]          H [C3H7]         H [C4H9]
```

Alanine	Valine	Leucine
$H_2N{-}CH(CH_3){-}COOH$	$H_2N{-}CH(C_3H_7){-}COOH$	$H_2N{-}CH(C_4H_9){-}COOH$

Figure 11
Some Examples of Common Amino Acids

The main difference is the hydrocarbon chain, as shown within the rectangles. Amino acids form one of the critical groups of physiological compounds. There are about 20 amino acids that the body needs to function. They all have a core unit of amine and carboxylic acid.

It would be worth explaining the nature and function of a body cell wall. It is represented by a thick line. However, it is almost like a brick wall made of lipids.

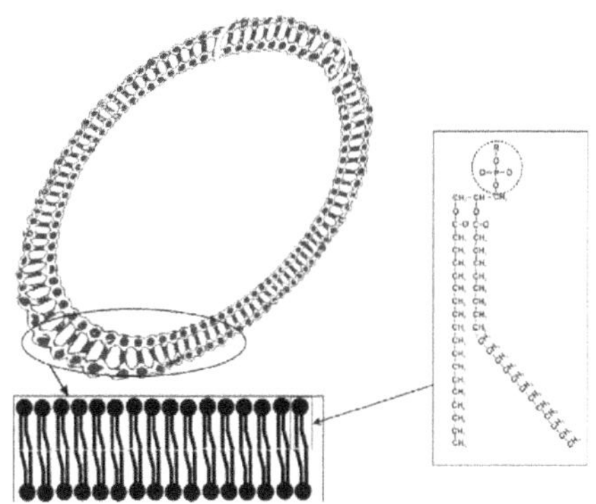

Figure 12
Schematic Representation of Cell Membrane

Figure 12's right side is built on propanol, as shown in Figure 5 (three carbons below the circle). Two carbon are attached to two fatty acids and one carbon to the phosphate group (in the circle). Considering that such compounds contain lipids (fats) and phosphorous compounds, they are called phospholipids.

When these phospholipids are exposed to water, they self-assemble into two-layered sheets with the hydrophobic tails (non-polar) pointing toward the center of the sheet. This arrangement results in two "leaflets," each a single molecular layer. The center of this bilayer contains almost no water. Instead, the assembly process is driven by interactions between hydrophobic (non-polar) molecules.

Everything that moves through the cell has to pass through this barrier, following fundamental chemistry principles. Thus, both externally exposed groups are polar in nature, held with the external aqueous milieu, while internal layers are "joined" with fatty tails. Compounds with fatty (non-polar) in nature will be sucked through this bilayer (in and out).

A commercial example of a cell bilayer is the liposome. Liposomes are small artificial vesicles of spherical shape that can be created from cholesterol and natural, non-toxic phospholipids. Due to their size and hydrophobic and hydrophilic character (besides biocompatibility), liposomes are promising systems for drug delivery. These liposomes (called nono particles) are being used to deliver mRNA-based vaccines.[11]

Biological membranes typically include several types of molecules other than phospholipids, mostly proteins (explained below). These protein molecules help the transfer of polar molecules to pass through. Often the protein molecules form as channels that allow the molecules to pass through, which the lipid layer would resist or not be allowed to pass through. Once the molecules pass through the cell membrane, they are processed by chemical reactions to fulfill the body/cell functions.

---

[11] https://pubs.acs.org/doi/10.1021/acsnano.1c04996

Next in line is another class of chemical compounds that are physiologically important. But before explaining further, so far, the chemical compounds with a straight chain are described. However, not all chemical molecules are straight-chain types. They are circular or loop molecules and are often named with the prefix cyclo. For example, cyclobutane (four carbon), cyclo-pentane (five-carbon), cyclohexane (six-carbon), etc. Such compounds are represented as follows:

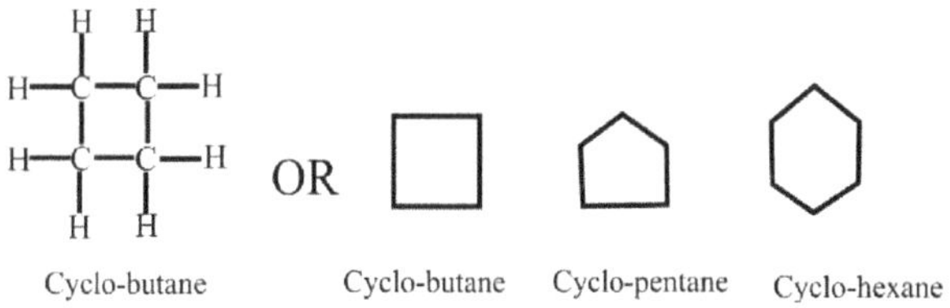

H H
H—C—C—H
H—C—C—H
H H
OR
Cyclo-butane Cyclo-butane Cyclo-pentane Cyclo-hexane

Figure 13
Loop or Cyclic Structures

In practice, carbon atoms are represented by a corner or an end; for example, butane becomes a square, pentane becomes a pentagon, and hexane becomes a hexagon. Hydrogen atoms are there as well but not represented with symbols.

On the other hand, their symbols represent all non-carbon and hydrogen atoms. Following are some examples of cyclic molecules with oxygen and nitrogen atoms.

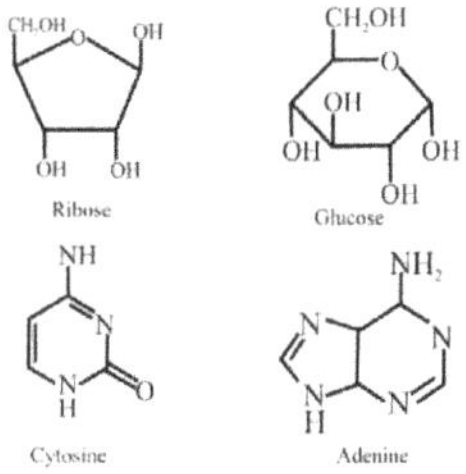

CH_2OH O OH
OH OH
Ribose
CH_2OH O OH OH OH OH
Glucose
NH N N O H
Cytosine
NH_2 N N N N H
Adenine

Figure 14
Some Examples of the Cyclic Structure of Physiological Importance

So far, what is presented are names and descriptions in chemical language. It might not be necessary for the general public to go into further detail. However, it is essential for anyone who works with chemical compounds, like anyone dealing with medicines, to understand how these chemicals are named and what characteristics one should expect. There is no need to memorize them. All one is required is to search on Google or read from an introductory chemistry textbook to have the needed information about the chemical names and structures and their spatial structure orientation.

For non-technical use, common names should work fine, e.g., glucose, sugar, salt. These chemicals are readily available from commercial suppliers in wholesale or retail variety with different purities depending on the need.

A simple analogy that comes to mind is the representation of a nut and bolt. A non-technical person needs to know the name of the part and its use and possibly the part number if it is to be used in an appliance. However, for development and manufacturing purposes, one would undoubtedly need details about material and technical drawings to make the product fit for the purpose and reproducible.

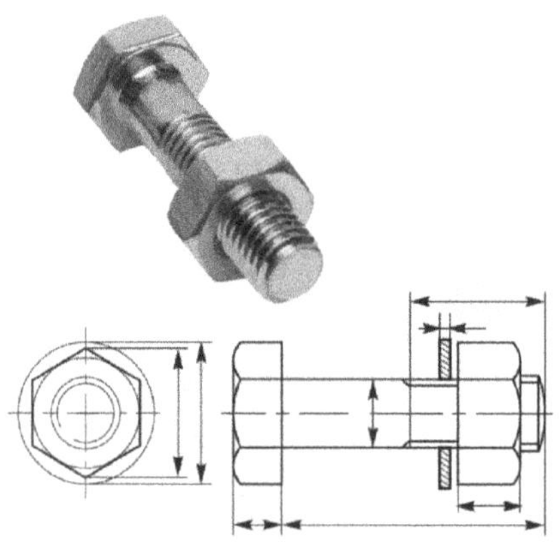

Figure 15
A Technical Drawing of a Nut and Bolt Pair Exemplifies Chemical Molecule Structure (Technical Drawing)

Therefore, in this respect, it should become evident that if anyone would work in medicines, such as their development and manufacturing, one must thoroughly be knowledgeable in details (e.g., chemical names, formulas, composition, structures, reaction, isolation, characterization, purification). In addition, one must explain clearly to the user, experienced or not in the field, information on the meaning of the formulas and their interactions within bodies based on well-established principles of chemical science.

So far, descriptions have been provided for single and multi-element molecules. However, in nature, many items are built on multiple units of molecules, i.e., repeatedly joining the same or different molecules in sequence, like carbon atoms in hydrocarbons.

For example, starch is a polymeric carbohydrate with numerous glucose units (cyclic six carbon-based units). Most green plants produce this polysaccharide for energy storage. It is the most common carbohydrate in human diets and is contained in large amounts in staple foods like wheat, potatoes, corn, and rice.

The general structure of starch is given in Figure 16. Note the repeat of the unit (glucose) could be anywhere from 300 to 600 times.

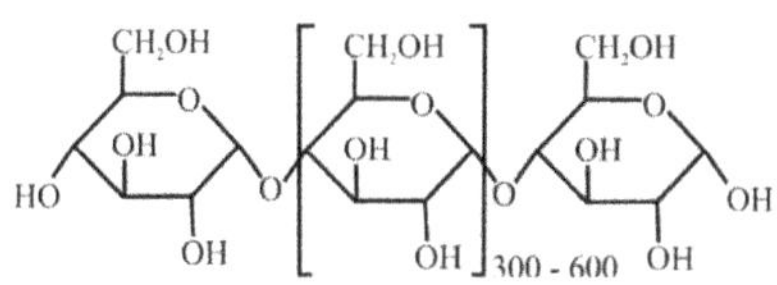

Figure 16
A Representative Chemical Structure of Starch

However, there are others where the repetition of multiple molecules forms such macromolecules. A common and popular example is DNA (DeoxyriboNucleic Acid) or RNA (RiboNucleic Acid) molecules, where four different types of units are repeated. These are also extremely long-chain molecules with four repeating units in any imaginable order. Therefore, when one describes that one is doing DNA/RNA sequencing, one determines the sequences of these four molecules. A daunting task, but it could be done and led to the award of the Nobel Prize to Dr. Kary Mullis. The structural formulas of the five commonly referred to molecules in this regard are shown below:

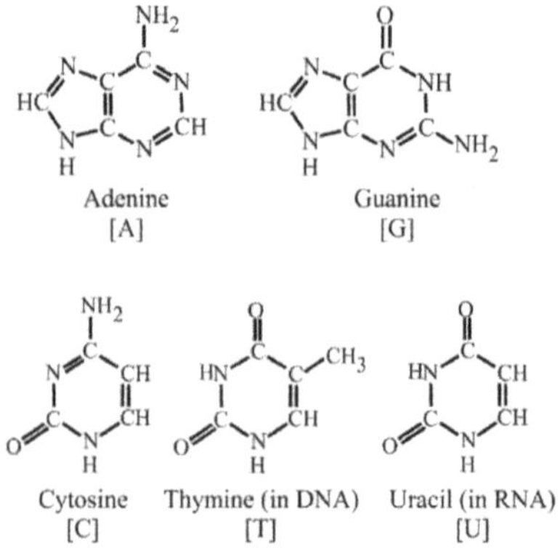

Adenine [A] Guanine [G]

Cytosine [C] Thymine (in DNA) [T] Uracil (in RNA) [U]

Figure 17
The Structural Formula of the Five Commonly Referred to DNA/RNA Molecules with their Alphabet Symbols in Parentheses

These are considered base molecules, but in DNA or RNA, they come with two more molecules, a sugar (pentose, ribose or deoxyribose) and a phosphate molecule, as shown below.

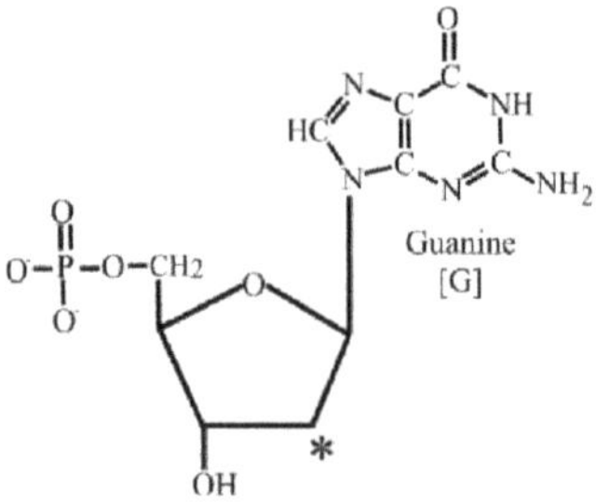

Guanine [G]

Figure 18
Structures of DNA with an -OH group at * called RNA

Joining these molecules together results in the commonly known structure of DNA or RNA, as shown here:

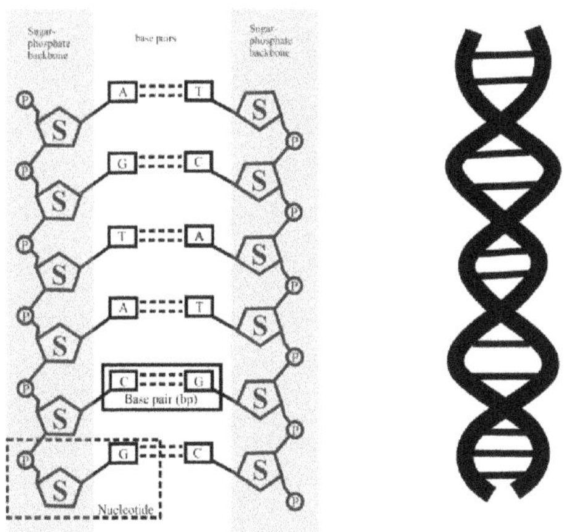

Figure 19
A Pair of a Double Strand of DNA
S=Sugar Molecule, P=Phosphate Molecules, and other Nucleic Acid Molecules and their Pairing

Another typical example of a biological long-chain molecule is proteins, in which amino acids are joined together to form large molecules. To understand protein structures and characteristics, one needs to understand their bonding characteristics.

```
      R1 O             R2 O             R3 O
      |  ‖             |  ‖             |  ‖
H-N-C-C-O-H     H-N-C-C-O-H     H-N-C-C-O-H
  | |               | |               | |
  H H               H H               H H

          R1  O    R2  O    R3  O
          |   ‖    |   ‖    |   ‖
      H-N-C-C-N-C-C-N-C-C-O-H
        | |   | |   | |
        H H   H H   H H
```

Figure 20
Amino Acids (with a Peptide Bond) are joined to Form Proteins

Amino acids are joined together by removing the hydroxyl (-OH) group of carboxylic acid and one hydrogen atom from the amino group forming a new bond called a peptide bond. The joining of three such amino acids with peptide bonds is shown in Figure 20—the R_1, R_2, and R_3 could be the same or different and represent a protein formula and structure. Customarily, these are called peptides; however, they are called proteins when the chain becomes longer. The division between peptides and proteins is rather vague. However, just for standard description, if one deals with a structure of less than 15 amino acids, one may consider a peptide, but beyond it would be a protein.

Some of the bonds in molecular structures of proteins and RNA/DNA are shown with dotted lines, not solid lines. These are weaker bonds, often referred to as hydrogen bonds, which help form the shape (helix) of the macromolecules.

From physiological perspectives, these protein molecules are built on about twenty different units (R), i.e., amino acids. Therefore, one can imagine the variety of combinations and permutations with large repetition. Indeed physiology is filled with this variety.

Human albumin is a small globular protein consisting of a single chain of 585 amino acids. It has a molecular weight of about 66500 units, compared to hydrogen and carbon atoms with 1 and 12 units, respectively.

Often pictures of proteins are shown in images of colored ribbons. However, scientifically studying them requires focusing on the chemical composition, often ignored in the literature.

In general, proteins are critical molecules essential for all living organisms. By dry weight, proteins are considered largest unit of cells. Proteins are involved in virtually all cell functions. A different type of protein is devoted to each role, with tasks ranging from general cellular support to cell signaling and locomotion. Proteins are classified into different types; antibodies and spike proteins (cell proteins) are the most commonly mentioned in describing the COVID-19 virus.

Some other chemical terms to be remembered in explaining the biological/physiological phenomenon are described below.

Polar and non-polar compounds: In a broader sense, a non-polar molecule, steroid (e.g., corticosteroid), is the one that will dissolve in non-polar solvents such as alcohol, oil, fat, and polar molecules such as common salt, are the ones that are soluble in water.

Absorption: a term often used to describe the transfer of nutrients or medicines into the bloodstream following their ingestion or application to the body. Consider that one has a thin hollow tube made of (candle) wax and a flowing solution containing some non-polar compound. The non-polar molecule will be attracted (stick) to the surface of the tube. More molecules will stick to the surface, pushing through to the other side of the tube. This passage of molecules through the layer (membrane) is called absorption. It is the most common mechanism of transferring molecules from one part of the body to another, including nutrient absorption from the GI tract into the bloodstream. Almost all medicines are absorbed into the body/bloodstream through this process. The long tube of the GI tract represents such a waxy tube to bring the drug or nutrients into the bloodstream or body through the absorption mechanism.

Acidity and Basicity of Molecules

Like polar and non-polar properties of molecules, in general, molecules and their environments are classified as acidic, neutral, or basic. A number on the pH scale defines acidity or basicity. A pH scale is like a temperature scale on a thermometer. The temperature scale is defined based on water (0°C for frozen water while 100°C for boiling water). Between 0 and 100 are equally divided into 100 bars or divisions. In the case of pH, the scale ranges from 1 to 14. One (1) means highly acidic, while 14 is least acidic or the most basic, seven neutral. As a reference, purified water is considered neutral, i.e., having a pH of 7. One thing to keep in mind is that, unlike the temperature scale, which is linear, the pH scale is exponential. i.e., each number represents a power of 10. i.e., if pH changes from 1 to 2, it is not double but an increase of 10 times. In the body, the pH spread from 1 to 3 (stomach), 5 to 7 (small intestine), and 7 to 8 in large (large intestines).

This combination of temperature, acidity, basicity, and polarity of molecules and their environment plays a critical role in body maintenance (homeostasis), drug efficacy, and safety.

Ions and Ionization

Considers putting salt in water. One does not realize it as it is not visible to the naked eye, but the molecule (sodium chloride, or NaCl) splits into two parts, i.e., sodium-ion and chloride-ion represented as Na^+ and Cl^-, respectively. This process is natural and is called ionization or ion formation. Most medicines are non-polar but are supplied in salt form for easier patient handling and storage.

Chemical vs. Biological Reactions

A chemical reaction is a process in which one or more substances, the reactants, are converted to one or more different substances, the products. Substances can be elements or compounds. The simplest example of a reaction would be combining two hydrogen molecules with one oxygen molecule and adding a spark. In chemical terminology, it is represented as:

$$2H_2 + O_2 \longrightarrow 2H_2O$$

Figure 21
Example of a Chemical Reaction and its Representation

Another example would be of formation of margarine, as mentioned earlier. The carbon-carbon double bonds of oils are filled (saturated) with hydrogen atoms through hydrogenation.

Usually, such hydrogenation causes a higher oil melting point (i.e., margarine). The hydrogenation or saturation of oil will result in the solid or semi-solid paste at room temperature for easier spread. The chemical reaction may be represented as follows.

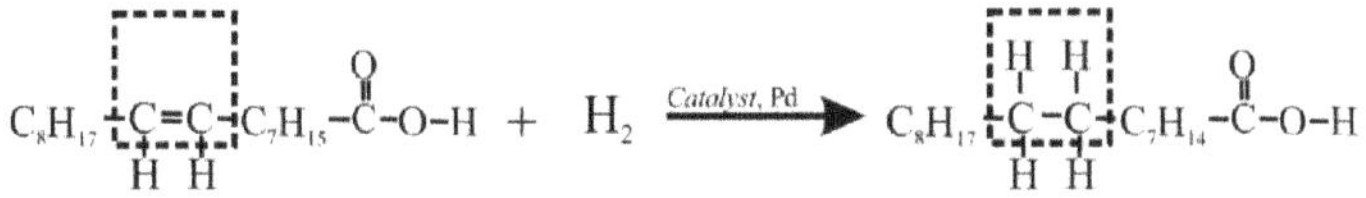

$$C_8H_{17}-\underset{H}{C}=\underset{H}{C}-C_7H_{15}-\overset{O}{\overset{\|}{C}}-O-H \;+\; H_2 \xrightarrow{\text{Catalyst, Pd}} C_8H_{17}-\underset{H}{\overset{H}{C}}-\underset{H}{\overset{H}{C}}-C_7H_{14}-\overset{O}{\overset{\|}{C}}-O-H$$

Oleic Acid

Oleic Acid
(hydrogenated or saturated)

Figure 22
Example of Hydrogenation for Converting a Double Bond into a Single Bond by Adding a Hydrogen Molecule

Extensive chemical reactions occur in the body. However, the body uses natural catalysts, often specific to a molecule or a class of molecules (also called substrates). The catalysts are called enzymes (mostly protein-based), and their name ends with "ase." So, for example, the enzymes which work with lipids (fat) are called lipases. The ones that work with protein are called proteases. Concerning viruses, PCR (Polymerase Chain Reaction) is commonly mentioned where polymerase is an enzyme that causes the polymerization or polymerization reaction.

The point of the above description is to elaborate that people should be very clear that the names described in literature originate and belong to the subject of chemical science. Therefore, a clear understanding of the subject of chemistry, its terminology, and laws and principles must be described or used in medical and/or pharmaceutical areas.

Chapter 6—Post-Normal Science

POST-NORMAL SCIENCE is a concept developed by Silvio Funtowicz and Jerome Ravetz who attempted to devise a methodology of inquiry appropriate for contemporary conditions.

The typical case is when "facts are uncertain, values in dispute, stakes high and decisions urgent."

In such circumstances, we have an inversion of the traditional distinction between hard, objective scientific facts, and soft subjective values. Now we have value-driven policy decisions that are hard in various ways, for which the scientific inputs are irremediably soft.

In a pressure situation, with media hype over the supposed global threat of a terrifying pandemic, solutions were rushed through. Rather than taking the normal 10 years to bring to market a safe and effective vaccine, drug manufacturers and policymakers went all in on post-normalism and paid scant attention to empirical verification.[12]

The above is a clear example of how objective science is made subjective, to fit an agenda by the trickery of turning science into the cult of the *expert* who no longer has to be robust in evidence collection, validation and replication, but who merely forms a group of enough peers (usually fellow university academics) to be able to declare a consensus of opinion to help their pay-master politicians determine major policy decisions. In essence, this invites corruption such that the politicians cherry-pick the *scientists* who will be most willing to concoct *secret science* to bolster a narrative.

Such experts are the least trustworthy.

Let us take, for example, Professor Neil Ferguson of Imperial College, London, who is empowered by the political class to seemingly make outlandish predictions of doom and gloom of mass deaths from a potential new strain of coronavirus. Failure to be in

---

[12] https://en.wikipedia.org/wiki/Post-normal_science

any way reliable with your prognostications has no negative impact on your career or credibility once you are among the inner circle of approved government experts. We know this because Ferguson's career as a government expert is a litany of failed predictions.

As Graeme McMillan wrote in May 2020:

> *I have no doubts that Professor Neil Ferguson is an extremely clever man.*
>
> *He is highly influential in all aspects of Health Policy. He is an advisor to the WHO, the European Union, and has had the ear of every UK Prime Minister since 2001.*
>
> *Governments around the world use Imperial research to underpin policy decisions, and their staff are among the world's most influential scientists.*
>
> *And yet, Prof. Ferguson remains an enigma. He has the unfortunate habit of making outrageous claims about the severity of new viral outbreaks throughout his career.*[13]

As McMillan explains, with the Creutzfeldt-Jakob Disease (CJD) hysteria in 2001, Ferguson predicted 136,000 deaths in the coming decades. But actual CJD deaths by 2015 amounted to 226.[14]

Ferguson provided policymakers with a range of mortality between 50 and 50,000. During the H5N1 outbreak in 2005 doomsayer Ferguson predicted 200 million deaths. H5N1 actual total was 455.

In 2009, Prof. Ferguson was quoted as saying (regarding H1N1):

---

[13] https://principia-scientific.com/discredited-origin-of-covid-19-lockdown-computer-models/

[14] Ibid.

> *This virus...is likely to spread around the world in the next six to nine months and when it does so it will affect about one-third of the world's population...*

That was another way-off bit of guesswork, but despite his penchant for outlandish scaremongering this expert remains the golden boy of the UK political class.

This nightmarish outcome for government science was predicted in 1961 by the outgoing US President. In his farewell speech (1/17/1961) President Eisenhower raised the issue of the Cold War and role of the U.S. armed forces. He described the Cold War:

> *We face a hostile ideology global in scope, atheistic in character, ruthless in purpose, and insidious in method...*

Eisenhower warned about what he saw as unjustified government spending proposals and advised:

> *...we must guard against the acquisition of unwarranted influence whether, sought or unsought, by the military-industrial complex.*

In the current case, it is the medical-industrial complex, working with eugenicist co-conspirators among the billionaire class, who are driving this post-normal science around COVID-19.

The medical-industrial complex was a term first coined in the 1971 book, The American Health Empire (Ehrenreich and Ehrenreich, 1971) by Health-PAC. The medical-industrial complex is composed of the multibillion-dollar enterprises including doctors, hospitals, nursing homes, insurance companies, drug manufacturers, hospital supply and equipment companies, real estate and construction businesses, health consulting, accounting, and banks.

The New England Journal of Medicine has an excellent paper on this phenomenon, titled *The New Medical-Industrial Complex* by Arnold S. Relman M.D. published back in 1980.

Relman said:

> *This, in barest outline, is the present shape and scope of the "new medical-industrial complex, a vast array of investor-owned businesses supplying health services for profit.... The new health-care industry is not only very large, but it is also expanding rapidly and is highly profitable. New businesses seem to be springing up all the time, and those already in the field are diversifying as quickly as new opportunities for profit can be identified. Given the expansive nature of the health-care market and the increasing role of new technology, such opportunities are not hard to find.*[15]

Writing for Principia Scientific International in 2021, William Walter Kay determined:

> *...the "complex" has become the standard mode of structuring the planning and control activities of corporate banking, agribusiness and mass communications.*[16]

Kay warned that complexes are sprawling Big Government/Big Business alliances wherein cross-penetration blurs public-private lines. Complexes are broader, deeper and more flexible than

---

[15] https://www.nejm.org/doi/full/10.1056/NEJM198010233031703

[16] https://principia-scientific.com/covid-19-and-the-global-medical-industrial-complex/

cartels. They decentralise to the point of being leaderless: The web is spidery but there is no single spider.

Kay notes:

> *Covid-19 is to the global medical-industrial-complex what the Iraq War was to the US military-industrial-complex.*

Covid-19 is a bonanza for many, especially Big Pharma.

In truly complex fashion, the US Government's Operation Warp Speed allocated $10 billion to vaccine R&D. Pfizer received $2.5 billion for R&D and was further aided by National Institute of Health technology transfers and clinical trials. Pfizer's partner, BioNtech, developed its mRNA vaccine with $440 million from the German Government. Pfizer received another $2 billion from Uncle Sam to retrofit factories.

Moderna's mRNA vaccine was developed by NIAID and the Biomedical Advanced R&D Authority. According to the New York Times:

> *The mRNA vaccines in which we are now staking so much hope wouldn't exist without public support through every step of the development. (Oxford professors developed AstraZeneca's Covid vaccine.)*

With such evidence, we feel justified in demanding independent peer reviewed (perhaps third party review is more appropriate) studies, not funded by multinational pharmaceutical companies. All the peer-reviewed studies of short-term safety and short-term efficacy have been funded, organised, coordinated, and supported by these for-profit corporations; and none of the study data have been made public or available to researchers who don't work for these companies.

It is our contention that for the past one hundred years,

science has become overly invested in (virtual or imaginative) modelling rather than practical, real-world testing and evaluation of physical phenomena, the reactions and processes that are proven reproducible in nature.

We lament the over-reliance on computer modelling where mistakes made in the inputting of data and algorithms that presumptive reality cause a multiplicity of modelling errors because of society's creeping over-reliance on computers to do the thinking and analysing for us.

Chapter 7—Chemistry/Science for Isolation, Purification and Characterization of Substances

ISOLATION AND CHARACTERIZATION of compounds is perhaps the most critical aspect of understanding medicines' development and use.

Yet, it may be the least described and understood in the medical and healthcare professions.

In reality, no illness or medication should be defined or developed without a thorough understanding of this topic; however, this is hardly the case. It could also be argued, and explained later in the section, that the entire COVID-19 pandemic resulted from the misunderstanding of this subject, i.e., isolation and characterization of the virus and the medicines/vaccines.

Isolation of molecules forms the backbone of disease identification and product development. The isolation process may be explained as follows.

Extraction (or sample preparation) is the first step for isolating compounds irrespective of their origin, plant, animal, or dirt (e.g., ores). Primarily, the material is crushed and homogenized with a solvent such as aqueous (water or buffers) or organic (non-polar) solvents such as hexane.

In most cases, the extraction is conducted with liquid-liquid or solid-liquid mixing. Liquid and ground solids are shaken with a liquid where molecules of interest have a higher affinity (solubility) and transfer to that phase.

Complete extraction of the compound of interest requires multiple extraction steps.

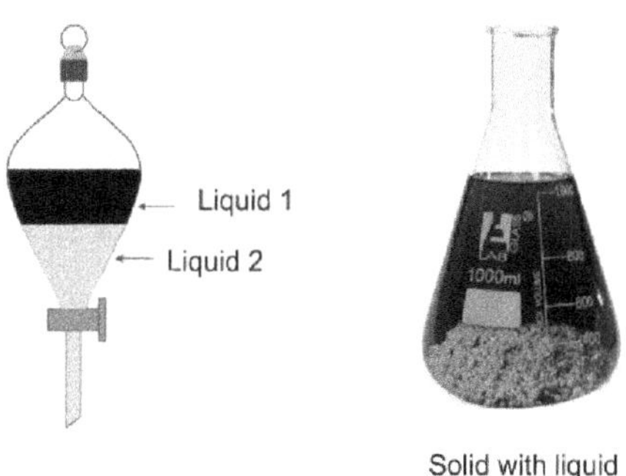

Figure 23
Schematic Representation of the Most Basic Isolation Step, i.e., Extraction

The liquid where the molecules are expected is decanted or passed through a filter to remove debris. This step is usually repeated several times to obtain maximum content.

One of the most commonly used techniques for separation and isolation is chromatography. The chromatographic separation process is explained in Figure 24 below.

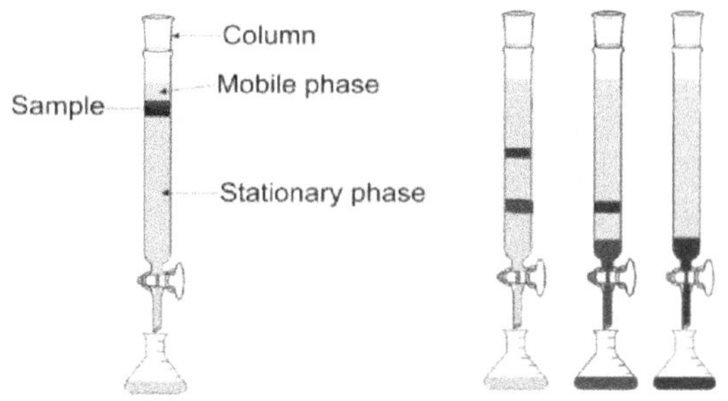

Figure 24
Schematic Representation of the Chromatographic Technique

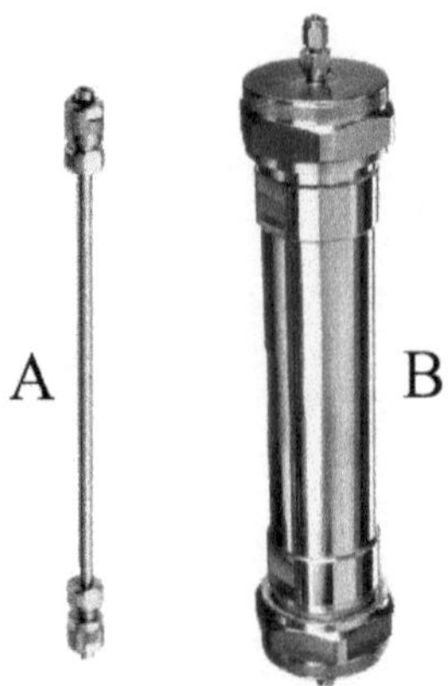

Figure 25
Typical Liquid Cchromatography Columns, A= for Analytical/Testing, B=Commercial Scale for Isolation/Extraction

Figures 24 and 25 show examples of chromatographic columns containing solid powder (stationary phase) while liquid or solvent (mobile phase) is run through it from top to bottom. A sample of interest is placed on the top, which moves down through the column with the mobile phase. Components of interest will move downward at different speeds depending on their affinity to the stationary and mobile phases. This differential mobility causes the separation of the molecules, and compounds will come out of the column one after another.

The process is the same as with a separatory flask or test tube. However, working with the column approach is far less messy, requires minimal material, and provides exceptionally high efficiency.

Column chromatography could provide multiple extraction steps (called plates) in hundreds and thousands.

The number of plates increases by decreasing the particle size of the stationary phase (increasing the surface area). However, doing so requires higher pressure (sometimes very high, 200 to 300 atmospheric) to move the mobile phase through it. This resulted in metal columns instead of glass columns filled with very fine particles, often in the range of 3 to 10 μm, in stationary phases.

It is called liquid-solid chromatography or simply liquid chromatography. If the mobile phase is gas, it becomes gas-chromatography instead of liquid chromatography. Usually, in the latter case, gas flows through the liquid, often coated onto solid particles. Therefore, gas chromatography provides an even higher number of plates than liquid chromatography. The choice of chromatography depends on the nature of the analyte. For example, gas chromatography should be used if the analytes are gaseous or liquid with low boiling temperatures (perfumes, oils, etc.).

However, if the molecules are solid, then liquid chromatography is used. Other chromatographic techniques are paper and thin layer chromatography and electrophoresis. However, most separations in modern-day are conducted using liquid and gas chromatography.

A typical outcome of chromatography is a graph called a chromatogram representing the number of components exiting the column detected using various monitoring systems such as ultraviolet-visible (UV) spectroscopy.

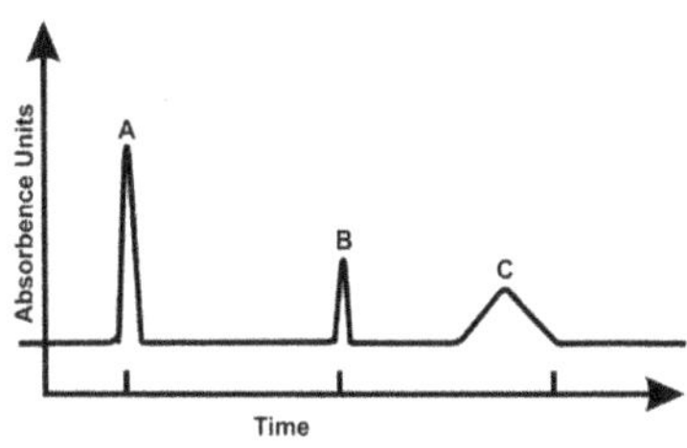

Figure 26
A Typical Output Record of Chromatographic Separation

It cannot be emphasized enough that chromatography is a vital part of any separation and isolation of the substance of interest.

Therefore, some understanding of the subject is critical, in fact, a must. If not used or tried (such as in the case of viruses or

PCR testing), then there should be cause for concern or suspicion.

Assuming the peak represents a single or pure component leads to a characterization step. First, the compounds from the eluting peaks are collected, followed by concentration by evaporation, freeze-drying, etc., as considered appropriate so the analyst has some solid or liquid material in hand to work with.

Here the characterization means establishing the compound's identity, starting with the physical description of the material, such as color, melting point, and elemental composition. Then, the most critical one is structure elucidation. There are many techniques available to achieve the characterization part. It is like putting together puzzle pieces to build a picture.

Another way of describing it is whether a compound falls in one of the categories described earlier, such as fat, carbohydrate, protein, DNA, etc., or a completely new one, often rarely.

The three most commonly used techniques are infrared (IR) and ultraviolet (UV) spectroscopy, nuclear magnetic resonance (NMR), and mass spectrometry. In essence, these techniques help determine the compound's structure and spatial arrangements of the atoms.

For IR spectroscopy, the molecule produces some characteristic pattern (fingerprinting) used to establish the identity or no identity.

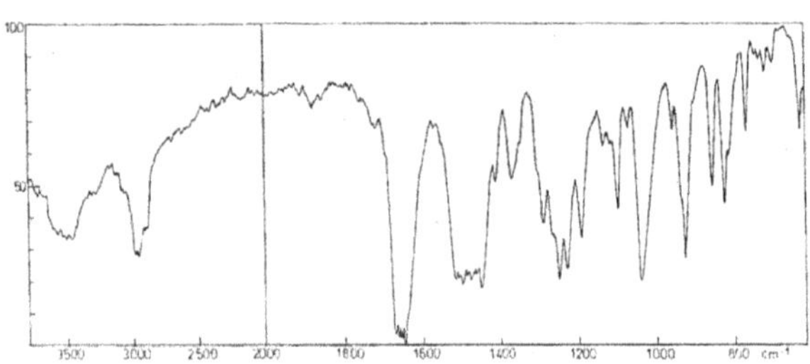

Figure 27—A Typical IR Spectrum[17]

[17] Saeed Qureshi, 1980, *A Chromatographic Investigation of Pepper Alkaloids.* Ph.D. Thesis, University of Ghent, Belgium

The NMR spectrometry works on the principle that when molecules are placed in a strong magnetic field, the nuclei of some atoms will begin to behave like tiny magnets. If a broad spectrum of frequency waves is applied to the sample, the nuclei will start resonating at their own specific frequencies.

The resonant frequencies of the nuclei are then measured and recorded as an NMR spectrum.

By analyzing such peaks, one can then determine the 3D structure of molecules and observe how they move.

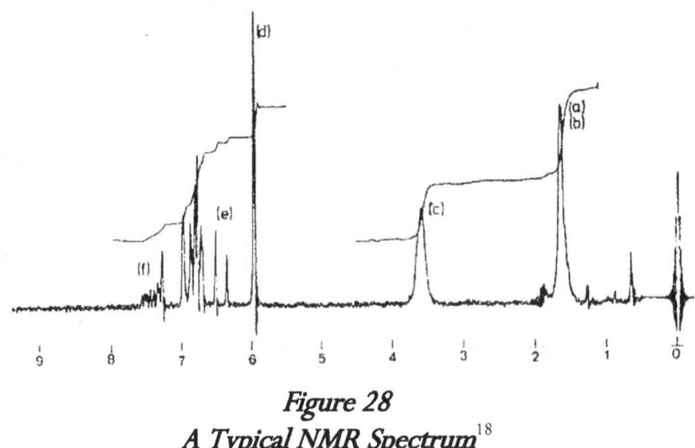

Figure 28
A Typical NMR Spectrum[18]

Perhaps the most powerful technique in this regard is mass spectrometry. It works on the principle that if a compound is passed through under very controlled temperature and pressure conditions and bombarded with an electron beam that would split the molecules into many small pieces reproducibly.

This analytical technique is used to measure the mass-to-charge ratio of ions. The results are presented as a mass spectrum, a plot of intensity as a function of the mass-to-charge ratio. Mass Spectrometry (MS) is used in many fields and applied to pure samples and complex mixtures.

---

[18] Ibid.

In a typical MS procedure, a sample, which may be solid, liquid, or gaseous, is ionized. This ionization causes sample molecules to break up into positively charged fragments or become positively charged without breakup. The fragments are then separated according to their mass-to-charge ratio. Next, the ions are detected with a charge multiplier. Results are displayed as spectra of the signal intensity of detected ions vs. the mass-to-charge ratio. The atoms or molecules in the sample can be identified by correlating known masses (e.g., an entire molecule) to the identified masses or through a characteristic fragmentation pattern.

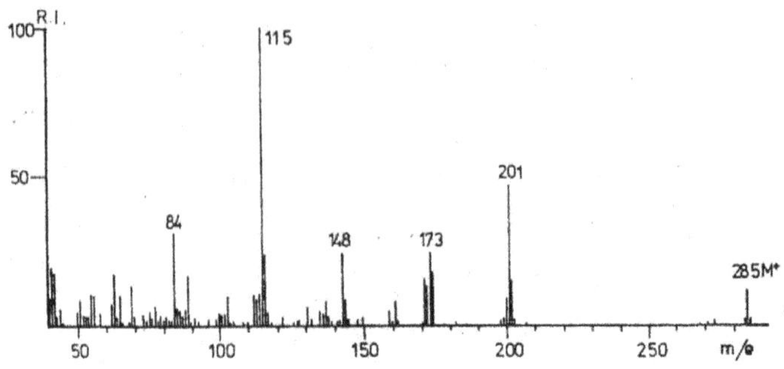

Figure 29
A Typical Mass Spectrum[19]

Structural libraries are built on such fragmentation that helps recognize compounds that would give the same or similar pattern. It is important to note that this technique can analyze extremely small amounts of compounds, often in the picogram (10^{-12}) range.

Two techniques such as chromatography and mass spectrometry—coupled together, called hyphenated techniques, and are frequently used, i.e., separation and characterization in one step.

---

[19] Ibid.

Although these testing techniques provide the almost true nature of the compound, to confirm the identity positively, still, as a final step compound of interest is synthesized or isolated from a different source. This isolated compound goes through all these tests to confirm its identity and characterization. Hence such a compound becomes the reference standard for future test developments and evaluation of samples for compounds of interest.

The bottom line is that at the end of the isolation and separation steps, one would have pure compounds or particles (RNA or virus) in a vial or test tube. However, the literature does not provide an example of a pure isolated virus specimen. But claims are made that the virus has been isolated. So, why is this discrepancy? A virus isolation claim is simply a false claim.

Let us explore how the described chemistry background and isolation processes (chromatography as an example) could be applied to isolate a virus specimen.

First, what is a virus? A virus is a particle (virion) consisting of an outer protein shell and an inner core nucleic acid. Virions, such as coronavirus, presumably also have surface spikes known as spike protein (or S-protein).

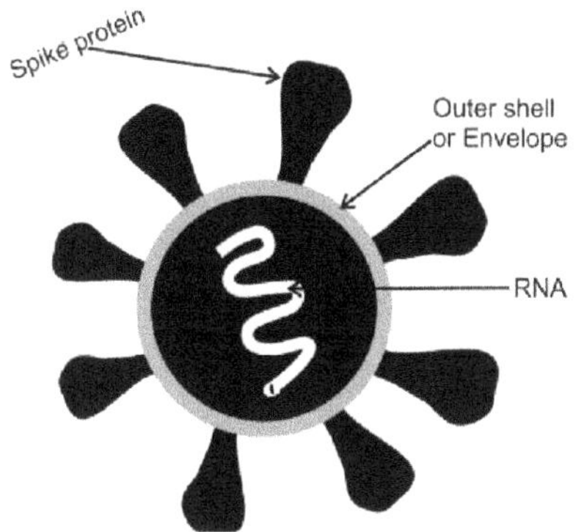

Figure 30
A Typical (Coronavirus) Virus Particle with the Most Commonly Referred Components

So, the particles must be isolated if someone likes to study the virus's characteristics and composition and its link to an illness or developing treatment (such as vaccines).

Before the isolation step starts, one must have this virus's source. It is commonly assumed that virus particles reside in the nose and mouth and move in and out through aerosols and vapor droplets. Therefore, one needs to collect a large quantity of exhaled air from human subjects.

Three options in this regard may be used (1) by collecting used face masks (e.g., collected in dumpsters) at large public places such as malls and supper stores and extract viruses from them; (2) installing large filters with suction to collect exhaled air from the customers. These could be good options available during a widespread pandemic like the one we are observing now.

The third possibility is to grow the virus by inoculating the cell culture with human swab samples. This is often described as the only option under non-pandemic situations.

The soiled masks and filters may be soaked in water, buffer, or organic solvent to bring the virus into the liquid part. The cell-culture media could be used as such. Next, the liquid portion is passed through high-grade filters to remove debris, followed by a liquid-liquid extraction to bring the virus component into smaller liquid portions. Finally, this extracted portion of the raw source is applied to a commercial-scale chromatographic column for isolating or separating the virus. First, the eluent from the chromatographic column should be collected in fractions. The fractions would be monitored for virus with the help of a pre-determined test, commonly a bio-assay. Sometimes one chromatographic separation might not provide pure enough virus fractions. Therefore, different stationary and mobile phases might require multiple chromatographic separation steps. However, one round of chromatographic separation should be often sufficient to isolate the virus.

Once the appropriate viral fraction is identified with the bioassay, it should be dried out by evaporating or freeze-drying the solvent. This specimen will now be ready for characterization and identification, such as elemental analysis and structural determination using methods described above such as (IR, UV, visible spectrophotometric, Nuclear Magnetic Resonance (NMR), and mass spectrometric). This will result in the isolated and pure virus specimen and become the virus's reference (gold) standard.

This pure and isolated virus must be further "crushed" to break down into components like spike proteins and RNA and be further separated and fractionated using chromatographic columns. Once the pure fractions of each component, i.e., RNA, spike proteins, and envelop proteins, are obtained, they are again characterized individually using the earlier techniques. However, here considering the nature (polymers) of the compounds, i.e., proteins and RNA, are to be further characterized using sequence analysis to identify and establish their long chains and three-dimensional structures appropriately.

Once these are fully characterized, they will become the reference (gold) standards of the respective RNA and proteins and may be used as potential markers for the virus. Note that the unique nature of the compound RNA and protein will mark them as unique viruses and variants.

These specimens would be free from impurities and available as viruses, variants, and their respective RNAs and proteins. This is a common and standard approach used in modern-day laboratories to separate and isolate molecules and compounds.

Indeed, the presented is an overview of the technique and process. However, the specifics may be different from laboratory to laboratory. For example, the choice of extraction solvents, chromatographic columns, mobile phases, virus monitoring approach, etc., would vary. Still, the outcome has to be exactly the same as a pure (unadulterated or free from other components) virus, variant, DNA/RNA, or spike protein sample.

Let us now explore how the medical and virology professions describe the isolation of viruses and see if they indeed have isolated the viruses.

Medical and virology experts claim that the virus has been isolated, its genome sequenced, and specimens commercially available. So why is a discrepancy?

The following would help clarify the confusion by critically evaluating the most commonly described procedure for isolating the virus in the medical community.

The procedure is explained with the help of a study described by the CDC,[20] considered the most authentic and authoritative source in this regard. The title of the publication is *Severe Acute Respiratory Syndrome Coronavirus 2 from Patient with Coronavirus Disease*, United States, i.e., finding the coronavirus from a patient.

Further, the article describes:

> *In this article, we describe* ***isolation*** *of SARS-CoV-2 from a patient who had coronavirus disease (COVID-19) in the United States and described its genomic sequence and replication characteristics.*

So, there should not be any doubt that the article is about the virus's isolation (i.e., extraction or obtaining) and then characterizing its genomic sequence.

From the publication's Methods section:

> *Nasopharyngeal (NP) and oropharyngeal (OP) swab specimens were collected on day 3 postsymptom onset, placed in 2-3 mL of viral transport medium, used for molecular diagnosis, and frozen. Confirmed PCR-positive specimens were aliquoted and refrozen until virus isolation was initiated.*

Explanation: sampling was conducted and confirmed based on PCR

---

[20] https://wwwnc.cdc.gov/eid/article/26/6/20-0516_article

testing, which can confirm nothing, as explained later.

Further explanation: Viral Transport Medium is suitable for transporting samples suspected to contain a virus. It is a ready-to-use solution containing Hanks balanced salt solution, heat-inactivated FBS (Fetal Bovine Serum), gentamicin sulfate, and amphotericin B.[21]

From the publication's Cell Culture, Limiting Dilution, and Virus Isolation section:

> *We used Vero CCL-81 cells for isolation and initial passage. We cultured Vero E6, Vero CCL-81, HUH 7.0, 293T, A549, and EFKB3 cells in Dulbecco minimal essential medium (DMEM) supplemented with heat-inactivated fetal bovine serum (5% or 10%) and antibiotics/antimycotics (GIBCO. We used both NP and OP swab specimens for virus isolation.*

Explanation: Vero cells are cells derived from an African green monkey's kidney and are one of the more commonly used mammalian continuous cell lines in microbiology and molecular and cell biology research.[22]

Further explanation: Fetal Bovine Serum (FBS) is the liquid fraction remaining after the blood drawn from the bovine fetus coagulates. Cells, coagulation fibrinogens, and proteins are

---

21 https://www.thermofisher.com/order/catalog/product/A48750BA?ef_id=CjwKCAiArOqOBhBmEiwAsgeLmW_qz2zapzp0Dgldbe1GmmVhwDcmKv8GFgC49uDzFbjp2V3VHz715BoCfewQAvD_BwE:G:s&s_kwcid=AL!3652!3!481196734962!e!!g!!viral%20transport%20medium&cid=bid_clb_cce_r01_co_cp0000_pjt0000_bid00000_0se_gaw_nt_pur_con&gclid=CjwKCAiArOqOBhBmEiwAsgeLmW_qz2zapzp0Dgldbe1GmmVhwDcmKv8GFgC49uDzFbjp2V3VHz715BoCfewQAvD_BwE

22 https://www.ncbi.nlm.nih.gov/pmc/articles/PMC2657228/

removed through centrifugation to produce serum.[23]

Further explanation: Dulbecco minimal essential medium (DMEM): a mixture of about 30 ingredients.[24]

From the publication:

> *...we pipetted 50 μL of serum-free DMEM into columns 2-12 of a 96-well tissue culture plate, then pipetted 100 μL of clinical specimens into column 1 and serially diluted 2-fold across the plate.*
>
> *We then trypsinized and resuspended Vero cells in DMEM containing 10% fetal bovine serum, 2× penicillin/streptomycin, 2× antibiotics/antimycotics, and 2× amphotericin B at a concentration of 2.5 × 105 cells/mL. We added 100 μL of cell suspension directly to the clinical specimen dilutions and mixed gently by pipetting.*

Explanation: Trypsinization is a process of cell dissociation using trypsin, a proteolytic enzyme that breaks down proteins. When added to cell culture, trypsin breaks down the proteins, which enable the cells to adhere to the vessel. Trypsinization is often used to pass cells to a new vessel. When the trypsinization process is complete, the cells will be in suspension and appear rounded.

From the publication:

> *We then grew the inoculated cultures in a humidified 37°C incubator in an atmosphere of 5% CO2 and observed for cytopathic effects*

---

[23] https://www.thermofisher.com/ca/en/home/references/gibco-cell-culture-basics/cell-culture-environment/culture-media/fbs-basics/what-is-fetal-bovine-serum.html

[24] https://himedialabs.com/TD/AT068.pdf

> *(CPEs) daily.*

Explanation: Cytopathic Effect (CPE) is structural changes in a host cell resulting from presumed viral infection. CPE occurs when the infecting virus causes lysis (dissolution) of the host cell or when the cell dies without lysis because of its inability to reproduce.[25]

From the publication:

> *We used standard plaque assays for SARS-CoV-2*

Explanation: Plaque assays measure infectious SARS-CoV-2 by quantifying the plaques formed in cell culture upon infection with serial dilutions of a virus specimen.[26]

From the publication:

> *When CPEs were observed, we scraped cell monolayers with the back of a pipette tip. We used 50 µL of viral lysate for total nucleic acid extraction for confirmatory testing and sequencing.*

At the end of the experiment or isolation step, they state:

> *We used 50 µL of viral lysate for total* ***nucleic acid extraction.***

But, the experiment was conducted to extract or isolate the virus. Where is the extracted virus? Nowhere! The virus has never been extracted or isolated.

The publication clearly shows that the authors seriously lack an

---

[25] https://www.britannica.com/science/cytopathic-effect

[26] https://currentprotocols.onlinelibrary.wiley.com/doi/full/10.1002/cpmc.105

understanding of isolation science. Therefore, all the claims made by the authors and the represented organizations regarding virus isolation must immediately be withdrawn. Further, a thorough audit of the so-called science at the medical organizations (CDC) should, in this regard, be initiated urgently by some authority competent in isolating and purifying materials.

Responses to Ms. Christine Massey have recently confirmed the lack of isolation of the virus received from 164 institutions worldwide, including the CDC, stating that they do not have any sample of pure isolated virus.[27] [28]

Furthermore, it is often fiercely debated that:

> *There are numerous examples of scientists isolating SARS-CoV-2, the virus that causes COVID-19, and sequencing its genome. The argument about purification relates to 19th Century microbiological theory that does not apply to viruses. The novel coronavirus has been proven to exist and has caused millions of deaths worldwide.*

This quote is from Reuters Fact Check[29] and is a blatant scientific lie. In reality, it is repeating the lie of the CDC to obtain a US patent (US 7,220,852 B1) for coronavirus. The title of the patent is *CORONA VIRUS ISOLATED FROM HUMANS.*

It states as follows:

---

[27] https://www.fluoridefreepeel.ca/fois-reveal-that-health-science-institutions-around-the-world-have-no-record-of-sars-cov-2-isolation-purification/

[28] https://drive.google.com/file/d/1xkRLVli4ZVvZ-j-Cw3OTABqvwGijueY2/view?fbclid=IwAR1HhXDT1J7cg6tmMa1VyXjci9Ivr20nPYT1w7j1gMvIg6G0P4Ggknr4EGc

[29] Fact Check—SARS-CoV-2 has been isolated, and its complete genome has been sequenced—https://www.reuters.com/article/factcheck-covid-rna-idUSL1N2LS27P

> *SUMMARY OF THE DISCLOSURE: A newly isolated human coronavirus has been* ***identified*** *as the causative agent of SARS, and is termed SARS-CoV. The nucleic acid sequence of the SARS-CoV genome and the amino acid sequences of the SARS-CoV open reading frames are provided herein.* [emphasis added]

The title says isolation, but the description says *identified* as the causative agent, not a virus—and is an inaccurate and deceptive statement.

> *This disclosure provides methods and compositions useful in detecting the presence of a SARS-CoV nucleic acid in a sample and/or diagnosing a SARS-CoV infection in a Subject. Also provided are methods and compositions useful in detecting the presence of a SARS-CoV antigen or antibody in a sample and/or diagnosing a SARS-CoV infection in a Subject.*

It provides a method for detecting the nucleic acid and presence of an antigen, *not* the virus.

> *IV. SARS-CoV Nucleotide and Amino Acid Sequences The current disclosure provides an isolated SARS-CoV genome, isolated SARS-CoV polypeptides, and isolated nucleic acid molecules encoding the same.*

This is the isolation of genome, polypeptide and nucleic acid, not the virus, but chemical molecules/compounds. Even here, no evidence or record is provided for a physical sample of the chemicals mentioned.

Another patent US 7,897,744 B2, Assignee (The Public Health

Agency of Canada) is titled *SARS Virus Nucleotide and Amino Acid Sequences and Uses Thereof.*

Under Summary of Invention, it states:

> *...the invention provides a substantially pure SARS virus nucleic acid molecule or fragment thereof, for example, a genomic RNA or DNA, cDNA, synthetic DNA, or mRNA molecule.*
>
> *In some cases, an isolated nucleic acid molecule is intended to mean the genome of an organism such as a virus.*
>
> *DETAILED DESCRIPTION OF THE INVENTION: In general, the invention provides nucleic acid molecules, polypeptides, and other reagents derived from a SARS virus,...**not the virus.***[30]

There is no mention of the virus or isolating the virus. Instead, they describe the chemical components. Again, the patent deceptively implies and falsely promotes an imaginary and presumed virus.

A few words about the ultracentrifugation technique which is often described in the medical literature regarding virus separation or isolation and purification.

Centrifugation is a simple technique for separating components in a liquid solution.

For example, if one shakes some sand particles with water in a test tube and leaves them on the table, the sand particles will quickly settle at the bottom of the test tube.

However, if the particles are small and light, they may not settle at the bottom of the tubes quickly and often remains suspended in water. The spinning of the test tubes is done to pull

---

30 https://patentimages.storage.googleapis.com/71/8a/e4/9ddf2dccfa4554/US7897744.pdf

the gelatinous material down with centrifugal force using centrifugation machines. Therefore, it is another technique for separating, like filtering the content (solid/liquid) in a flask or test tube.

The ultracentrifugation technique uses extremely high spinning speeds and gravitational force (g) in thousands of multiples. Such a force pulls the extremely small particles, even macromolecules such as proteins, DNA/RNA, etc., towards the bottom o the tubes.

A variation of this technique, i.e. gradient ultracentrifugation, is conducted where the aqueous phase is replaced by stacking different concentrations of sugar (sucrose) solutions in the test tube.

Applying the test sample (treated swab sample) at the top and then spinning the tubes results in spreading particles across based on particles' size and density. The band (position) where the specific types of particles settle is pipetted out and used for identification and characterization.

Due to the heterogeneity in biological particles, this differential centrifugation suffers from contamination and poor recoveries. From the virus isolation perspective, gradient ultracentrifugation's separation or spreading ability does not isolate the virus particles but anything in a certain size and density range. Therefore, it still provides a mixture of the original sample, perhaps less dense and more homogenous than the applied sample. The particles' mass, density, and shape dictate the position of the (particles) band.

How did this range become established as no virus so far, has been isolated?

However, knowing where the virus will fall when centrifuged requires the specimen of the virus for calibration purposes which, as noted above, is not available at present. Therefore, one cannot determine which band will contain the presumed virus.

One often sees the electron microscope pictures of the isolates with pointed arrows showing the particles considered viruses. This is purely imaginative or guesswork because there is no evidence that these are virus particles, let alone the coronavirus. Electron microscopic pictures like Figure 31 do not identify anything. It is like taking pictures of a wash after washing the clothes and pointing some dirt particles in it as gold because they appear yellow.

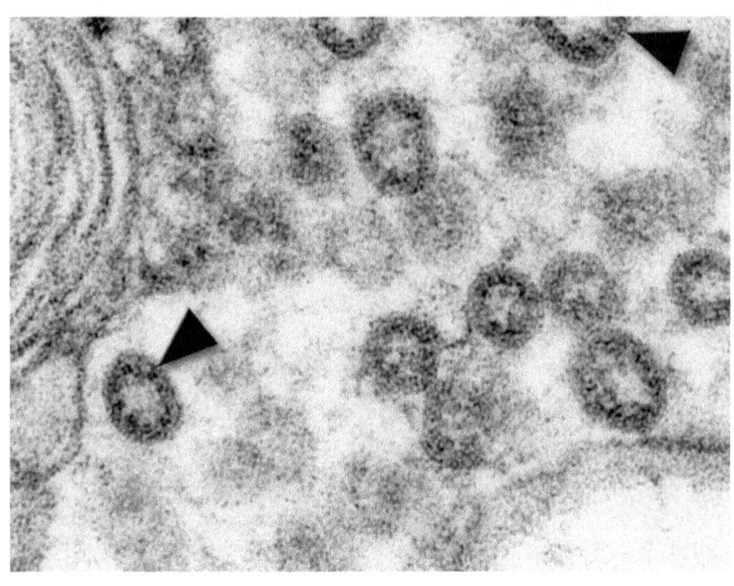

Figure 31
Transmission electron microscope from a tissue culture isolate reveals a number of severe acute respiratory (SARS) virus particles.[31]

The isolation (extraction) and purification of particles of the viruses and their identification and characterization still are missing. They are needed before the existence of the viruses is claimed and confirmed.

This shortcoming may be explained differently, that virologists confuse the culturing step with an isolation step, which is incorrect. Instead, a culturing step is to grow or produce something in a

[31] https://www.cdc.gov/about/history/sars/index.htm

broth. The broth is often even messier than the starting material or mixture.

Therefore, it can never be considered as an isolation or purification step, which comes after and is entirely different and separate from the culturing step.

Chapter 8—Chemistry/Science of Analysis and Testing

THE BEDROCK OF analysis and testing is the availability of a purified and fully characterized reference standard of the item which the test is supposed to test—in this case, the virus and RNA. Once one has the reference standards available, as described in an example above, then and only then can one start working with analytical methods or test development. Scientifically speaking, it is impossible to develop a test without the availability of a reference standard.

An example of a standard test development procedure using the chromatographic technique is described in a paper called *Determination of B_6 Vitamers in Serum by Simple Isocratic High-Performance Liquid Chromatography.*[32]

Vitamin B_6 is one of the B vitamins and an essential nutrient. The term refers to six chemically similar compounds, i.e., "vitamers," which can be interconverted in biological systems. Its active form, pyridoxal 5 -phosphate, serves as a coenzyme in many enzyme reactions in amino acid, glucose, and lipid metabolism.

Extraction Procedure: 500 µl of the serum sample was mixed with 1.0 ml perchloric acid in water (8%) to denature and precipitate the proteins. This mixture was vortexed for one minute, then centrifuged at 2500 RPM at 4 °C for 5 minutes. Next, the supernatants (from two repeat extractions) were applied to an adsorption material, followed by elution with 8.0 ml of 0.1 M citric acid in the water, discarding the first 2.0 mL. This sample would become concentrated or extracted plasma containing B_6 vitamers and other impurities and would require further separation to identify and quantify vitamin B_6.

---

[32] Saeed A. Qureshi & Hide Huang, Journal of Liquid Chromatography, 13:1, 191-201, (1990)

As explained earlier, a High-Pressure (performance) Liquid-solid Chromatographic (HPLC) technique was used for separation and qualification. The separation output (or chromatogram) is shown below from a spike water solution of vitamin B_6 and its closely related components (vitamers). The peak numbers reflect individual vitamers and their chemical names and structures in Figure 33.

Note that these are not different forms of vitamin B, but B_6 only demonstrates the separating capability of modern separating techniques.

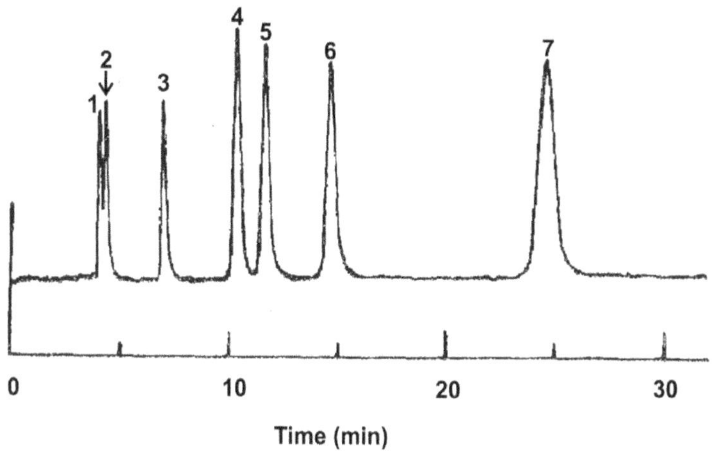

Figure 32
A Chromatographic Separation of B6 Vitamers using Chromatographic (HPLC) Technique. Numbers Indicate Individual Isomers as Noted in Figure 33

Once the separating capability of the chromatography is established, spiked plasma (with reference standards) is extracted using the extraction procedure and injected onto the column. Chromatograms of spike and blank plasma are shown in the Figure below, demonstrating free from interference in the areas of vitamer elutions.

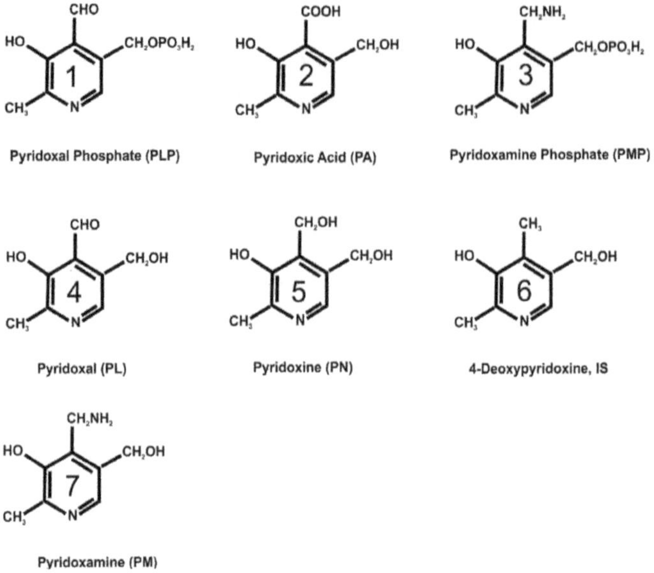

Pyridoxal Phosphate (PLP) | Pyridoxic Acid (PA) | Pyridoxamine Phosphate (PMP)

Pyridoxal (PL) | Pyridoxine (PN) | 4-Deoxypyridoxine, IS

Pyridoxamine (PM)

Figure 33
Names and Chemical Structures of Vitamers B6 where Numbers Identify the Peaks in the Chromatograms Shown in Figure 32

Once it is established that the individual vitamers can be separated, the linearity of their response is established to confirm that the method can measure the vitamers quantitatively.

Next, the method's accuracy is established by extracting known amounts of spike vitamers (reference/gold standards) and comparing their peaks (area or height) to reference peaks indicating that the method can recover the most amounts present. A higher of 80 to 90%+ recovery is always desirable. Otherwise, one needs to find the issue and address it appropriately. Finally, the linearity and extraction recovery data are summarized in Table 1.

Another thing to note is that when an actual sample is analyzed, it is often spiked with a compound having similar chemical characteristics as the compound of interest but should elute on the chromatogram as a separate from the vitamers. The purpose of using this extra compound, called Internal Standard (IS) is to reflect that the whole process of extraction and chromatography has gone through as expected by recovering the anticipated amount of the internal standard.

The chromatogram shown is from method development and validation for dog plasma samples. In addition, the blood samples from rats, monkeys, and humans were also tested in the study to establish that the extraction and separation procedure fit for the purpose, i.e., the method has been validated for its intended use.

Generally, this is how a test or method development and validation is done, which forms the backbone of any method. However, additional steps may be used for higher specificity and sensitivity. But the fundamental principle remains the same and is a must requirement for scientific and regulatory purposes.

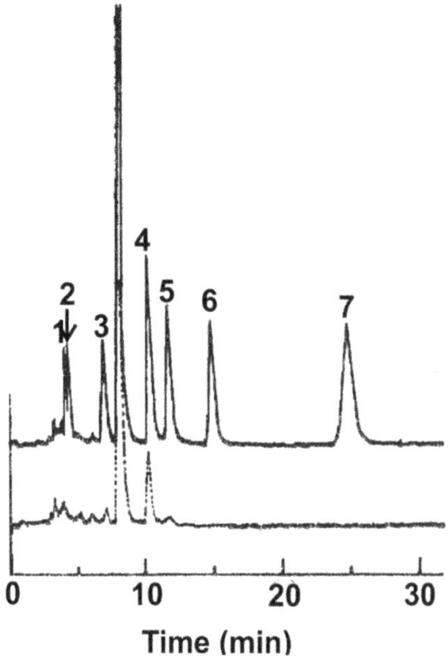

Figure 34
Chromatographic Separation of B6 Vitamers, Rat Blood Sample (below), and Spiked Blood Sample

Vitamer	Correlation Coefficient (r)	Extraction Recovery	
		Amount Added (ng)	%Recovery (± CV)
PLP	0.991	2400	91.43 ± 7.65
PA	0.995	480	95.62 ± 5.24
PMP	0.993	240	98.29 ± 5.93
PL	0.990	480	88.63 ± 1.77
PN	0.991	960	102.56 ± 5.67
PM	0.990	960	96.85 ± 6.69

Table 1
Correlation Coefficients of Concentration vs. Response

In general, a method validation report must provide an outcome of at least four parameters that the method is: (1) sensitive enough to detect the item it is supposed to detect; (2) repeatable or reproducible; (3) specific, i.e., it should able to see the item without the interference from other co-existing impurities; (4) a pure and certifiable reference product must be available. The critical aspect to note here is that if the reference or standard is unavailable, the other three items mentioned, i.e., specificity, sensitivity, and reproducibility, cannot be established. Thus, one cannot have a valid test or method.

Now let us consider the PCR test used to test the SARS-CoV-2 virus. It is to be noted that the PCR test is not a test for virus, illness, infection, or cause-of-death test. It is an RNA test that is assumed to be connected to the virus, illness, or infection. There is no evidence in the literature showing the RNA tested is from the virus. As noted above, establishing this requires extracting the virus sample for RNA.

This can only be shown as described under the virus isolation part above, i.e., first, the virus must be isolated, and then the RNA isolated from it and characterized. No such studies have ever been conducted; hence, no reference (gold) standard is also available for RNA. Therefore, developing a valid test or method for the RNA is impossible.

However, for the convenience of the readers, below is the summary version of the PCR test as currently employed.[33]

PCR (Polymerase Chain Reaction) is a method to analyze a short sequence of DNA (or RNA) in samples containing only minute quantities of DNA or RNA. PCR is used to reproduce (amplify) selected sections of DNA or RNA. Ingredients required for a PCR test.

- "Primers," short DNA sequences of the DNA to be copied.
- An enzyme (polymerase) for reading DNA code and assembling a copy.
- A pile of DNA building blocks (nucleotides) that the polymerase needs to make the DNA copy.
- These ingredients are added to the 50 µl of the lysate obtained from the "isolation" step, not the isolated virus.

Three steps are involved in a PCR to reproduce or amplify a selected section of the DNA. These three steps are repeated several times, called cycles. The cycles mean heating and cooling of the content of the reaction mixture. Amplification takes place in three steps:

1. **Denaturation:** At 94°C, the double-stranded DNA opens into two pieces of single-stranded DNA.
2. **Annealing:** At around 54°C, the primers pair up with the single-stranded "template" of the DNA.
3. **Extension:** At 72°C, the DNA nucleotides are coupled to the primer, making a double-stranded DNA molecule.

---

[33] MedicineNet, https://www.medicinenet.com/pcr_polymerase_chain_reaction/article.htm#what_is_pcr_polymerase_chain_reaction

The formation of DNA copies is monitored by measuring the fluorescence intensity of the light emitted from the polymerization reaction. This is the PCR test to monitor the RNA and, by extension, the virus.

The PCR is used to reproduce (amplify) DNA or RNA and is done by adding selected primers and building blocks (nucleotides) to duplicate the DNA/RNA. The selected primers decide which DNA/RNA to amplify. But, how does one decide which DNA or RNA represents the virus to amplify?

That is supposed to be the purpose of the (PCR) test—to find the target DNA/RNA. This cannot be done unless one has DNA/RNA reference standard from the virus to develop the primers. Analysts assume or decide which DNA/RNA to look for and, based on this assumption, choose the primers to use.

It means analysts do not see/test the DNA/RNA but "create" one of their likings, especially the new ones or novels. Furthermore, by changing the primer, one generates the variants per need.

In short, if and when scientists like to create a new virus and its variants, they will ask everyone to use a different set of primers. No one can determine any RNA of the virus from natural sources. The claimed RNAs derived from imaginative exercise-based PCR amplification and computer modeling. The chemical process only can determine the virus from natural sources based on extraction and isolation, as described above.

The CDC publication referenced above[34] described that 50 μl of lysate or isolate was used to extract the nucleic acid. This is another appalling misunderstanding and misrepresentation by medical professionals and experts that nucleic acid and virus are not the same things or interchangeable.

A nucleic acid is not a virus but supposedly part of the virus. However, even this assumption has not been proven correct because to be sure that nucleic acid is part of the virus, the virus

---

[34] https://wwwnc.cdc.gov/eid/article/26/6/20-0516_article

must be physically extracted or isolated. This has not been done yet. Therefore, working with nucleic acids (extracting, monitoring, or amplifying) to link them to virus presence or absence is a scientifically irrelevant and invalid claim.

In addition, as a pure isolated virus or its DNA/RNA (i.e., reference standard) is unavailable, it is impossible to validate the PCR test.

As described above, for validation (i.e., establishing sensitivity, specificity, and reproducibility of the test), one requires a reference standard of the virus RNA that is currently unavailable.

People inquire if the virus has not been isolated and purified; where do then the pictures of the coronavirus (ball with spikes) come from. As noted above, they are purely fictional and computer-generated based on the primers' selections and computer modeling/graphics.

The following analogy may explain the computerized process of generating a virus. Someone gets a sample of a couple of color-coded yarn threads and assumes they came from a sweater. Then, by duplicating millions and billions of copies of the pieces of yarn and joining them together, one builds a replica of the sweater.

But, how did the fabricator know that the yarn threads came from the sweater, not another type of cloth, scarf? Similarly, how would one establish that the amplified RNA is not from any of the originally present components (swab sample) or added for cell growth?

In fact, the virus might not even be present in the sample, and the RNA might be from a completely different or unknown source.

The nature of PCR can be explained with the help of a simple but very well-known chemical reaction called the Diels-Alder Reaction; let us abbreviate it as DAR.

Figure 35 shows that DAR is a chemical reaction to combine two chemicals having double bonds to create a cyclic or cyclo end product.

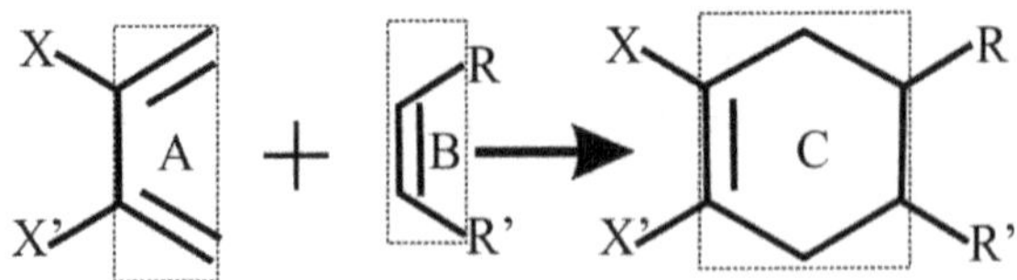

Figure 35
Diels-Alder Reaction

This chemical reaction between two molecules (A and B) produces a new chemical compound (C), a cyclic one. The created new molecule is dictated by the nature of the initial reactants or molecules. Let us assume the end product has a unique color as well. An analyst decides to use this reaction as a test to monitor the presence or absence of a single-double bond molecule (B) in a biological sample by adding a two-double bond molecule (A). Indeed, the reaction will occur, indicating the production of color (cyclic compound formation). However, it would not be possible to know the nature of the molecule produced (C) because its nature depends on the presence or absence of the R, R', X, and/or X' groups. Further, it would not be possible to know the source of the B, i.e. is it from the body's metabolic end product, something the person consumed, or perhaps impurity of the added reactants.

This is precisely the issue with the PCR test: reaction would occur by adding a chemical known as a primer (the equivalent of A in DAR). Still, one would not know the exact nature of RNA (the equivalent of B in DAR) present in the biological (swab) sample, its source, i.e., if it is from the body's debris, something the patient consumed, or just an environmental containment.

For a scientifically valid (PCR) test, the test must be applied after clearly isolating and characterizing the "B" component RNA from the virus, i.e., availability of the reference standard of the RNA of the virus of interest. As the virus's RNA is unavailable, the test becomes scientifically invalid or fraudulent. Hence all claims from such a test must be declared null and withdrawn.

The point is that the PCR test and its associated RNA

sequencing are irrelevant and/or invalid for monitoring the virus.

People often suggest the term "number of cycles," reflecting the number of repeats of the chemical reaction while conducting a PCR test. Regarding the accuracy of the PCR test, i.e., a higher number of cycles are perhaps causing the problem of showing a false positive. Hence, lowering of # of cycles may address the issue of false or higher positive results. Of course, not. The problem is not because of the number of cycles but the test itself. The test has never been validated, and it cannot monitor viruses—period.

So, in short, the PCR test has no meaning, and no one is currently monitoring the virus. It is critical to note: using and/or promoting non-validated tests is generally considered negligence and incompetence and could lead to third-party investigations with severe consequences. Recently a new test has become popular, called Rapid Test (or Antigen and Lateral Flow test).

What is a Rapid Test?

A Rapid Test is a quick test for COVID-19 that gives results in 30 minutes. The test kit is a hand-held device with an absorbent pad at one end and a reading window.

How is it different from a PCR test?

PCR tests detect the genetic material (RNA), while Rapid Tests detect proteins from the virus.

Also, although both tests require one to take a swab from the nose or throat, PCR tests must be sent to a laboratory, while Rapid tests can easily be done at home and give results in minutes.

How effective is the lateral flow test for COVID?

Data shows that Rapid tests are reliable for detecting the coronavirus. A recent study showed that when used correctly, Rapid Tests are likely to have a sensitivity above 80% and, in many cases, above 90%. It is important to note that the sensitivity and selectivity Rapid Test are measured against the PCR test, which has a serious problem, not against the virus or its protein. Like virus RNA, virus proteins have never been isolated.

Therefore, it is impossible to develop an appropriate and valid scientific Rapid Test.

Chapter 9—Criminal Global Cabal: Agenda 21 and COVID-19

A government big enough to give you everything you want, is strong enough to take everything you have.

WHAT MAY BE shown is that a criminal global cabal coordinated a societal 'Great Reset' to benefit themselves and usher in a collectivist technocracy of stark winners and losers. We show that the prime architects of this dystopia are the World Economic Forum, and their associates, implementing the UN's totalitarian Agenda 21.

For those with little or no awareness of this UN mission statement we urge you to search online. You will learn that Agenda 21 (Reinvented as Agenda 2030 and Agenda 2050) is a Plan to depopulate 95% of the world population by 2030. It is an action plan devised by the U.N. and signed by 178 governments. Its goal is the depopulation of humanity because "we are too many." [35]

Sir Walter Scott said, *"What a tangled web we weave when first we practice to deceive."*

For background information on the claimate change fraud, see Dr. Tim Ball's *Seven Ways to Spot Climate Change Propaganda.* [36]

For the past two generations, our children have been (falsely) taught to believe that the world is overpopulated, and we have reached a grave limit to natural resources. Agenda 21 has the goal of depopulation of humanity because *we are too many.* It is promoted by the elites as a way to *save the planet* and implemented by

---

[35] https://humansbefree.com/2019/03/agenda-21-reinvented-as-agenda-2030-and-agenda-2050-is-a-plan-to-depopulate-95-of-the-world-population-by-2030.html

[36] https://principia-scientific.com/dr-tim-ball-seven-ways-to-spot-climate-change-propaganda/

governments worldwide.

Bill Gates, prime vaccine promoter, even shared his view about how to achieve this goal by vaccinations and other means in a TedX lecture:

> *The world today has 6.8 billion people. That's heading up to about nine billion. Now if we do a really great job on new vaccines, health care & reproductive health services, we could lower that by perhaps 10 or 15 percent.*[37]

In this volume our focus is the pandemic, virus and vaccines misinformation.

> *A lie told once remains a lie, but a lie told a thousand times becomes the truth.*
> —Dr Joseph Goebbels

Agenda 21 is largely being implemented by NGOs funded by foreign countries via foreign aid budgets.

None of what is happening is about tackling a novel coronavirus.

Paul McGuire, an internationally recognized prophecy expert, speaker, minister, and author writes in his book The Babylon Code:

> *The true agenda of Agenda 21* [/2030] *is to establish a global government, global economic system, and global religion. When U.N. Secretary General Ban Ki-Moon spoke of 'a dream of a world of peace and dignity for all' this is no different than when the Communists promised the people a 'worker's paradise.'*

---

[37] http://humansarefree.com/2018/10/list-of-30-elites-that-support-and.html?fbclid=I

And the toxic experimental jabs and vaccine mandates appear to be doing their job in culling the witless sheeple (people). Two newly released studies show that—after a brief period of moderate protection—the experimental Covid-19 vaccines cause more sickness than they prevent.

The first study was released on MedRXiv by a team of researchers in Denmark.[38]

The study showed that those who received the Pfizer vaccine were 76.5% more likely to have a breakthrough infection than their unvaccinated counterparts once 90 days had passed—those who received Moderna's were 39.3% more likely.

The Canadian Covid Care Alliance—a non-profit government watchdog group of independent health care professionals—released a separate report that found similar conclusions. They reported:

> *...vaccine causes more harm than good.*
>
> *...The Pfizer 6 month data shows that Pfizer's COVID-19 inoculations cause more illness than they prevent. "Severe adverse events" were up by 75% in the trial group that had received the vaccine. Overall, adverse events that were attributed to the vaccine were an astounding 300% higher than in the placebo group.*[39]

Dr Robert Malone, who helped invent the core technology platform for the COVID-19 shots said:

> *Antibody enhancement is the vaccinologist's worst nightmare. That the product that you worked so hard to create actually causes worse disease.*

---

[38] https://www.medrxiv.org/content/10.1101/2021.12.20.21267966v2

[39] https://www.canadiancovidcarealliance.org/media-resources/the-pfizer-inoculations-for-covid-19-more-harm-than-good/

> *...We have the flawed clinical studies, I acknowledge that they're far from perfect and some might say designed to not be able to detect certain types of adverse events.*

Specifically, one thing they were not designed to detect is antibody dependent enhancement. And that's recognized by the FDA because in the FDA authorization letter for Emergency Use Authorization, they specifically say that the data are not sufficient to rule out antibody dependent enhancement; it remains a significant risk and suggested to the vaccine companies that they should do follow on clinical studies to detect the presence or absence of antibody dependent enhancement. But they didn't insist that the vaccine companies do that.

Dr Malone says the vaccine companies decided to take a pass and not investigate whether or not antibody dependent enhancement would occur.

Malone lamented,

> *What we do know is the distribution of virus levels, load, whether it's measured by PCR cycles or some measure that correlates more directly to titer, in the vaccinated is at least as high, those vaccinated that have 'breakthrough' infections, that become infected, is at least as high as in the unvaccinated and to my eye, it looks like there's suggestions that there may be a subset of patients that have been vaccinated and infected, that have even higher levels of viral replication than are present in the unvaccinated population. That, if that was to hold true, that would be the smoking gun demonstrating antibody dependent enhancement.*[40]

[40] https://www.bitchute.com/video/jHcDEhJgn3y6/

In an interview on October 13, 2021, Dr Malone went so far as to allege that regulatory agencies including the CDC and FDA are *profoundly corrupt* and have pushed experimental Covid-19 vaccines on the population with *grossly incomplete* data that does not meet even the bare minimum standards for safety.[41]

With that level of adverse events, it begs the question: are they trying to kill us? Well, yes, if we heed the words of the EU chief Ursula von der Leyen who told reporters:

> *I think, it is understandable and appropriate to lead this discussion now* [on] *how we can encourage and potentially think about mandatory vaccination within the European Union.*[42]

DRUG ADVERSE EVENT COMPARISON

FDA AND CDC DATA: WORLDWIDE

	Adverse events	Deaths	Deaths/year
1/1/1996 – 9/30/2021:			
Ivermectin	3,756	393	15
HCQ	23,355	1,770	69
Flu vaccines	197,816	2,001	77
Dexamethasone	83,599	15,910	618
Tylenol	112,244	26,356	1,024
Since 2020:			
Remdesivir	6,504	1,612	921
In 12 months:			
Covid vaccines	1,000,229	21,002	21,002

FDA FAERS system, CDC VAERS system. Reports from all locations worldwide. Data as of Dec. 24, 2021; downloaded Jan. 1, 2022.

The above chart was uploaded by U.S. Senator Ron Johnson to Twitter. It puts into context how much worse these experimental COVID jabs when comparing adverse events among the listed

---

[41] https://www.lifesitenews.com/news/744314/?utm_source=gab

[42] https://www.msn.com/en-gb/news/newslondon/eu-president-ursula-von-der-leyen-says-time-to-consider-mandatory-vaccinations/ar-AARlQdj

treatments. For his trouble, he got a warning label slapped on this tweet.

Publications like ours can be important for enabling wider understanding and ought to be read in tandem with the many articles and videos freely available elsewhere and at www.principia-scientific.org. But do not take our word on anything. Please do your own research.

You can check the Agenda 21 programs implemented in your city or town by searching Agenda 21 and the name of your city in any search engine.

At Principia Scientific International (PSI) you will see open and free discussion of every article, without censorship of reader comments, on important works. They include, but are not limited, contributors such as Dr Peter Breggin, Robert F. Kennedy, Paolo Bernard, Dr Steven Quay (*Origin of the Virus*), Dr Richard M. Fleming, Professor Christopher A. Shaw, Dr Zev Zelenko, Dr Vernon Coleman, Alex Berenson and numerous other hard-working experts, scientists, doctors, and investigative researcher/ journalists.

While what we present is disturbing, everyone who values their health and their freedom should make use of these resources.

Read avidly, do your own research and make up your own mind.

This book will hopefully help many be part of the growing resistance working to stop the insidious slide towards an oppressive global technocracy and mandatory vaccination which will fragment, divide and destroy humankind.

Chapter 10—Pandemic Panic Porn: Psyops against Humanity

IT IS NOT a pandemic—it is an IQ test. The Omicron variant (anagram of 'moronic') was a blunt clue to the millions who avoid cogitation upon the evidence revealed. At the time of writing, the Omicron iteration of this alleged new pathogen is widely agreed to be mild.

The folks who were told their lives would return to normal if they just complied. A coronavirus was called a cold before 2020.

Testing positive is not the same thing as being sick.

In South Africa, where that variant was first identified, the nation has opted for the lowest level of COVID-19 restrictions (level 1). But in London, a hot spot for pandemic panic porn, the UK government's 'expert' advisers at SAGE have been proven wrong every time. Yet their apocalyptic models are being parroted as truth by every media outlet.

Let us be clear on this: there is no SARS-CoV-2 pathogen,[43] and, as such, there is no Delta variant or Omicron variant. (It's all based on computer modelling rather than real-world observation.)

A study published in the journal *Social Psychology and Personality Science* tries hard to stigmatize those with the courage to stand up against the mob during coronavirus mania.

We can say this has been a psyops because government-funded psychologists are producing research[44] referring to individuals who resisted the mass hysteria resulting from the COVID-19 pandemic as mentally ill, claiming that they are more likely to exhibit psychopathic behavior.

43

https://war.jermwarfare.com/l/0lj0ynBLE23aKi9892pKIE8Q/cOeyE95aezYELvtqPhV55g/T76

44 https://www.psypost.org/2020/06/psychopathic-traits-linked-to-non-compliance-with

Psychological operations or PSYOPS are planned operations designed to influence emotions, motives, and objective reasoning among organizations, groups, and individuals. Any student of military history will tell you that wars don't just happen. PSYOPS are a routine function of governments at least since World War 2 when war time propaganda was deemed as being highly effective.

Operation Mockingbird, which we refer to elsewhere herein, has been the template for today's mind control of the masses via the mainstream media. The most successful operations come down to this simple strategy.

Piers Robinson, co-director of the Organization for Propaganda Studies,[45] is an internationally recognized expert on propaganda, but not just any propaganda. He explains that the most successful operations come down to this simple strategy: Identify your enemy, segregate, isolate, and then annihilate.

Robinson spoke at length about the propaganda aspects of COVID-19 in an August 04, 2021 interview with Asia Pacific Today of Australia. Robinson told interviewer, Mike Ryan:

> *COVID-19 is probably one of the biggest propaganda operations we've seen in history because of the global nature and the resources put into it. It was pretty clear from the beginning that propaganda was being employed.*[46]

Professor Robinson, a respected independent expert, identified how the British government used behavioral psychologists to exploit the media to manipulate public opinion by ratcheting up the level of fear surrounding the COVID-19 pandemic.

Robinson cited U.K. government documents from March 2020 referencing groups skilled in applied psychology to drive a

---

[45] http://propagandastudies.ac.uk/

[46] https://www.citizensjournal.us/propaganda-expert-warns-of-global-covid-psyop/

fear-based narrative causing behavioral change into making people follow the government's scheme.

One such group was SPI-B or the Independent Scientific Influenza Group on Behaviors. Robinson said:

> *The key part of SPI-B's behavioral change strategy that seems to have been adopted was to 'persuade through fear.'*

The persuasion section of the document states:

> *A substantial number of people still do not feel sufficiently personally threatened.*

Appendix B of the documents lists 10 options that can be used to increase social distancing in the public.

Option 2 advises:

> *Use media to increase sense of personal threat.*

SPI-B recommendations were to increase the sense of personal threat and "use hard hitting emotional messaging." This included taglines such as:

1. Anyone can get it. Anyone can spread it.
2. Don't put your friends and family in danger.
3. Stay home for your family. Don't put their lives in danger.
4. If you go out, you can spread it. People will die.

Use of hysterical media headlines was another big part of the strategy.

Big Tech quickly agreed to come on board with the censoring of all competing narratives that countered the fear-based approach. If that didn't work, the psychologists argued for *shame and approval* tactics. SPI-B psychologists knew that fear on its own would not persuade everyone. Messaging needed to be tailored to take into account different *motivational levers.* [47]

Outside the biased pronouncements of government *experts* those who are renowned independent experts, including Dr Paul Alexander, Dr Byram Bridle, Dr Geert Vanden Bossche, Prof. Dolores Cahill, and Drs. Sucharit Bhakdi, Ryan Cole, Richard Fleming, Robert W. Malone, Peter McCullough, Mark Trozzi, Michael Yeadon, Wolfgang Wodarg, and Vladimir Zelenko, among many others, consistently warn the world about the adverse effects resulting from COVID-19 experimental injections; they also warn about their long term effects. Such effects cannot be known at this time since most clinical trials will not be completed until 2023, and some as late as 2025.

The political class has nowhere to hide if called to account for their cavalier approach to public health dangers. The *safe and effective* false propaganda, put out by public officials who now are continuing to push this vaccine, is a clear breach of duty. A public office holder is subject to, and aware of, a duty to prevent death or serious injury that arises only by virtue of the functions of the public office.

Independents bemoan the insane obsession over testing everyone, regardless of whether they have any symptoms or even want to be tested.

And has anyone seriously asked the question: why does no one test positive for the old variants once the new one comes along?

Laura Dodson noted in her excellent book, *A State of Fear: How the UK Government Weaponised Fear During the Covid-19 Pandemic*, the whole pandemic may be summed up in one word: fear. Fear of a virus. Fear of death. Fear of losing our jobs, our democracy, our human connections, our health and our minds.

It's also about how the government weaponised our fear against us—supposedly in our best interests—until Britain became

---

[47] https://www.citizensjournal.us/propaganda-expert-warns-of-global-covid-psyop/

one of the most frightened countries in the world.

Out of the window went rational analysis as the fear took hold. It was a highly emotive period when families became divided, isolated, disorientated about what was real and what was not.

Add to the insanity the fact that almost the entire world had missed the crucial point that no lab had even isolated the SARS-CoV-19 virus yet. At the time of writing, there is no gold standard verification of the existence of this so-called deadly virus, which even the WHO concedes is no more a threat to us than any common influenza. Media fearmongers spawned this fake pandemic, not a microscopic pathogen.

Dr Vernon Coleman, Britain's most successful published medical doctor, explains in detail the reasons why the pandemic was an illusion based on faked data, abuse of elderly and sick patients, removal/suspension of essential care protocols, etc.

In the winter of 2021 Dr Coleman wrote:

> *The coronavirus is no more deadly than the flu-COVID-19 is merely the annual flu season rebranded as deadly pandemic. Doctors and hospitals have been bribed with a 'bonus' for listing ordinary flu deaths and COVID-19.*[48]

We recommend reading Dr Coleman's book *Endgame* which offers so much to think about. He was the first qualified medical practitioner in the UK to question the significance of the crisis now described as COVID-19, telling readers of his website www.vernoncoleman.com at the end of February 2020 that he felt that the team advising the Government had been unduly pessimistic and had exaggerated the danger of the virus.

At the beginning of March 2020, he explained how and why the mortality figures had been distorted. And on March 14th, he

---

[48] https://principia-scientific.com/dr-vernon-coleman-final-proof-the-pandemic-never-existed/

warned that the Government's policies would result in far more deaths than the disease itself.

Fortunately, Darwin's *survival of the fittest* is in full play.

Be your low IQ or high, when every waking moment of your day is flooded with lies and faux medical statistics and never any cogent counterpoints to offer balance, then you are on the slippery slope to surrendering any critical reasoning to Covidians (ostensibly, Malthusians) who spew vile hate to dissenters of their creed to *jab, jab, jab.*

The hate is real. The cultists use the term unvaccinated because it has negative connotations, with being unclean, unwelcome. This is to further demonise and degrade those of us who refuse to succumb to their propaganda. We, the authors of this book, prefer to describe ourselves as vaccine free and urge you to help change the narrative.

At least since 1919, western society has evolved towards being a technocracy with all the benefits and flaws that entails. Our centers of learning, schools, universities, teaching hospitals, etc. are only as good as their resources permit. And, as we shall see elsewhere in this book, resources are arbitrarily and capriciously directed towards a particular kind of knowledge which benefits those who own and run those institutions.

Revisionism of history and science in our universities is rampant and with the legacy media no longer doing real journalism, it seems there is no knowledge system now in operation that truly empowers the people. On the contrary, the vast body of human technical knowledge today far eclipses that of the generations before computers became ubiquitous.

As Dr Vernon Coleman laments in his excellent short book about the masks, *Proof That Face Masks Do More Harm than Good,* which everyone may read for free.[49]

---

[49] https://vernoncoleman.org/sites/default/files/2021-03/proof-that-face-masks-do-more-harm-than-good-v2.pdf

> *Politics and science merge in this realm.*
>
> *Power tends to corrupt and absolute power corrupts absolutely.*
> —Lord Acton

Principia Scientific International (PSI), among other independent STEM critics of the government technocracy correctly identified that those who are entrusted to create the computer models relied upon by the UN's Intergovernmental Panel on Climate and the World Health Organization are self-serving cronies of the elite.

The 'pandemic expert' relied on by many governments for dystopian lockdown policies that killed thousands of small and medium size businesses, Professor Neil Ferguson, was never held to account to justify his ludicrous computer model projections that millions would die if governments were not given full powers to shut down society. All premised on secret science.

The common denominator is always that these 'threats' to humanity come with a ready-made solution that involves stealing our freedoms and personal wealth to bloat the reach of governments and the coffers of the uber rich.

It is all done as if by magic with a few trusted insiders with their fingers on the keyboard programming the system and outputting solutions they will have you believe are objectively derived. Their mantra is *follow the science!*

> *There will be, in the next generation or so, a pharmacological method of making people love their servitude, and producing dictatorship without tears, so to speak, producing a kind of painless concentration camp for entire societies, so that people will in fact have their liberties taken away from them, but will rather enjoy it.*
> —Aldous Huxley

The goals of those in power are to preserve that power and to extinguish any threat to their hegemony.

At the start of the pandemic in 2020, everyone complied with government policies and were on the same side. The opposition became skeptical as the months elapsed and the reality of what was taking shape emerged. They were conservatively minded without being divided along party political lines.

Chapter 11—Secret Science

WE SAW THE first manifestations of COVID-19 secret science dirty tricks exposed over the safety and efficacy of hydroxychloroquine (+ zinc, Z-pac) as a reliable prophylaxis/treatment in March 2020.

At that time President Donald Trump announced he had:

> *...finished a two-week regimen of the anti-malarial drug hydroxychloroquine and is taking a victory lap, reminding those who opposed his taking of the drug that he survived it.*[50]

Of course, the legacy media retaliated by denouncing hydroxychloroquine as dangerous and Trump as a reckless, bigoted fool. Notwithstanding his later pronouncements that everyone gets the Big Pharma jabs, too. In an apparent flip-flop in December 2021 the former POTUS revealed he received a booster shot of the Covid-19 vaccine, drawing boos from a crowd of his supporters in Dallas.[51]

This left many skeptics of the Covidien cult wondering if Trump was still the populist leader in which they once placed so much of their trust. At the outset of the pandemic Trump had been unequivocal that both the pandemic and man-made global warming were part of the Chinese Communist Party to usurp America's pre-eminence as the world's leading power.

However, a study from June 2021 showed that Donald Trump was the main anti-vaccination influencer on Twitter in 2020.[52]

---

[50] https://www.westernjournal.com/trump-blows-media-water-epic-message-finishing-hydroxychloroquine-treatments/

[51] https://www.theguardian.com/us-news/2021/dec/20/trump-covid-19-booster-shot-crowd-boos

[52] https://www.psypost.org/2021/06/study-indicates-donald-trump-was-the-main-anti-vaccination-influencer-on-twitter-in-2020-61032

But regardless of Trump's true position, during the pandemic, we have seen in play the same tried and tested dirty tactics relied on for three decades by governments to fool us over the 'climate crisis.'

Whatever misgivings may arise about his position on the vaccines, Trump was consistent in asserting uptake of the jab was a personal decision and not the role for government to enforce.

The enduring message of the former occupant of the Whitehouse was that we must not trust the *fake news* which was in lock step in their Chicken Little cries about both climate and COVID-19.

With both these global disaster narratives the media playbook was the same: co-opt, bribe, threaten and coerce select scientists to concoct evidence to fit a government globalist UN narrative; rely heavily on computer models and ensure key data/coding is kept secret. Assure the public that the consensus view of the 'leading experts' supports the government narrative. Deploy a compliant media to act as gatekeepers so only the approved narrative is broadcast. Defame and de-platform critics (realists) and defy FOIA requests. Resort to the courts to chill dissent.

With western populations so ill-educated in the hard sciences, it was no wonder they became slaves to the Big Lie. Computer modelling by 'experts' is now akin to Truth.

Today we live in the era of narrow AIs (Artificial Intelligences) of ML (machine learning) agents and DL (Deep Learning) algorithms.

We call these new types of machines as automata, robots, bots, software, websites, search engine, algorithms, cars, computers, mobile phones, PCs, smartwatches and apps, all having some specific AI capacities. Note AI is based on the artificial—not on reality or truth.

The new machine age is characterized by specialized machines that relieve humans of all specific jobs, which machines do faster, cheaper and better than any humans do. Soon, it seems even our politicians will be replaced by them. At least that will help remove the façade that our elected officials are there to do our bidding.

The World Economic Forum, and their associates, implementing the UN's totalitarian Agenda 21 are the creators of the Great Reset socialist policy.

There is worrying evidence that the Great Reset, brought about during the pandemic, is not merely economic and culturally it is a reset of human biology itself. We learn that the virus pathogens we have been battling were patented years earlier by the likes of the Rothschilds and Microsoft billionaire, Bill Gates.

The Great Reset may fairly be described as the Marx communist manifesto in a new wrapper. From it, all capital and control become consolidated into the hands of the billionaire elites. The final step is the accomplishment of the abolition of all private property.

Michael Rectenwald, chief academic officer for American Scholars, wrote a compelling article about the Great Reset.[53]

In it, he explains that while Klaus Schwab, founder and executive chairman of the World Economic Forum (WEF), is credited with coining the term:

> *...the idea of the Great Reset goes back much further. It can be traced at least as far back as the inception of the WEF, originally founded as the European Management Forum, in 1971. In that same year, Schwab, an engineer and economist by training, published his first book, Modern Enterprise Management in Mechanical Engineering. It was in this book that Schwab first introduced the concept he would later call "stakeholder capitalism," arguing "that the management of a modern enterprise must serve not only shareholders but all stakeholders to achieve long-term growth and prosperity." Schwab*

[53] https://imprimis.hillsdale.edu/what-is-the-great-reset/

> *and the WEF have promoted the idea of stakeholder capitalism ever since. They can take credit for the stakeholder and public-private partnership rhetoric and policies embraced by governments, corporations, non-governmental organizations, and international governance bodies worldwide.*

Rectenwald specifically identifies that the Great Reset has its roots in climate alarmism.

He explains:

> *In addition to being promoted as a response to COVID, the Great Reset is promoted as a response to climate change. In 2017, the WEF published a paper entitled, "We Need to Reset the Global Operating System to achieve the [United Nations Sustainable Development Goals]." On June 13, 2019, the WEF signed a Memorandum of Understanding with the United Nations to form a partnership to advance the "UN 2030 Agenda for Sustainable Development." Shortly after that, the WEF published the "United Nations-World Economic Forum Strategic Partnership Framework for the 2030 Agenda," promising to help finance the UN's climate change agenda and committing the WEF to help the UN "meet the needs of the Fourth Industrial Revolution," including providing assets and expertise for "digital governance."* [54]

Rectenwald makes a strong argument that The Great Reset aims to usher in a bewildering economic amalgam—Schwab's stakeholder

---

[54] https://imprimis.hillsdale.edu/what-is-the-great-reset/

capitalism—which Rectenwald calls *corporate socialism* and Italian philosopher Giorgio Agamben has called *communist capitalism.*

At the root of both the man-made global warming catastrophe narrative and fearmongering over a fake virus from Wuhan is one of the WEF's many powerful *strategic partners,* BlackRock, Inc., the world's largest asset manager. BlackRock pretty much has a finger in every financial pie in the world's economic arena and is solidly behind the stakeholder model. In a 2021 letter to CEOs, BlackRock CEO Larry Fink declared that:

> *....climate risk is investment risk and the creation of sustainable index investments has enabled a massive acceleration of capital towards companies better prepared to address climate risk.*[55]

In their recent book on the Great Reset, Schwab and Malleret pit stakeholder capitalism against neoliberalism, and defined the latter as...

> *...a corpus of ideas and policies...favouring competition over solidarity, creative destruction over government intervention, and economic growth over social welfare.*

In other words, *neoliberalism* refers to the free enterprise system.

In opposing that system, stakeholder capitalism entails corporate cooperation with the state and vastly increased government intervention in the economy.

Or, to put it another way, the Great Reset is *capitalism with Chinese characteristics*—a two-tiered economy with profitable monopolies and the state on top and dismal socialism for the majority below.

---

[55] Ibid.

Chapter 12—COVID-19 Chimera

WE, THE AUTHORS of this book, are forthright in our view that COVID-19 was always a chimera, a sky dragon not unlike the false claims that carbon dioxide was a dangerous heat-trapping atmospheric gas. Junk science proliferates because the scientific journals are used as another corrupt extension of the lying mainstream media, which in turn also serve their corporate masters in the evolution towards the new two-tier economy Klaus Schwab and the globalist billionaire class envision for us all.

Schwab and Thierry Malleret wrote that if "the past five centuries in Europe and America" have taught us anything, it is that "acute crises contribute to boosting the power of the state. It's always been the case and there is no reason it should be different with the COVID-19 pandemic."

At the forefront of helping expose this new medical mythical foe has been David E. Martin, PhD. His expertise in this field is widely acknowledged.[56] He has contributed enormously to wider understanding of the origins of the 'deadly virus', also commonly referred to as a 'new corona virus', or as SARS-COV-2. Authorities tell us this is the proven cause of COVID-19 flu-like health problems.

In his presentation of November 5, 2021, at the Wise Traditions Conference, Dr Martin declared:

> *I am going to put my life energy on a single focus. And that is until we hold the criminal conspiracy accountable, and we end the emergency use authorization we're not done. And there is no other topic, right now, and I don't care what topic*

[56] https://www.davidmartin.world/

> *you love, there is no other topic that I am concerning myself with save the preservation of the next generation from a mass murdering, genocidal, psychopathic group of criminals who've decided that they can look at a five-year-old with contempt.*

In an interview by Stew Peters on July 20, 2021, David Martin set out the position that we at Principia Scientific International heartily endorse after Peters asked him if the virus was real.[57]

> **Peters:** *So, there really is no virus.*
> **Martin:** *No. this 'novel virus' thing…in 2002 Ralph Baric at the University of North Carolina [and others] file a patent on what they referred to as an infective, transmission-defective form of corona virus. Now think what I just said: …. What does that mean?*
>
> *In 1999 Anthony Fauci wanted to…use a virus as a vaccine. …That is actually a bad idea. But what he wanted to do was to take…the corona virus, and he wanted to make it mutated and chimerically altered and recombinated so that he could…target human lung epithelium… And what Ralph Baric's team did in 2002 they patented a whole bunch and variations of corona virus, which would make it more infectious to humans. ….*
>
> *When you make an infectious replication-defective virus that is meant to target human lungs so that it's more toxic…you built a Frankenstein! And in 2002 we knew that that Frankenstein was what it was. And remember that was the year before SARS number 1… We didn't have SARS*

[57] https://www.bitchute.com/video/6MS6wiOFb3dI/

until we made the weapon. ….

[Referring to media hype about -RS] *…where did the [so-called new corona virus-RS] come from is actually a total misdirect. …. The problem is nothing started in Wuhan. Nothing!*

Peters: *[So] the vaccine wasn't brought here to fight a virus, but the virus was here to introduce the vaccine.*

Martin: *Correct!*

It is no coincidence that these deadly viruses seem to have been patented.

Under patent law biological organisms cannot be patented. So, what gives here? It seems that we are blurring the boundaries and playing God with the very existence of humanity.

A major clue that the 'vaccines' are not the answer, but are really the problem, is the fact that governments have announced that if you don't have follow up boosters you will lose your vaccination status. Meanwhile, in locations such as Florida, which abandoned the idea of mask or vaccine mandates, has a 2.3% COVID positivity rate. And have you noticed that the word influenza has disappeared from common usage. Everyone who used to get colds and flu now has COVID.

Back in the UK, after booster-jabbing 57% of the population, Britain has recorded the highest number of cases since March 2020. But a little joined up thinking here helps the blinkers fall from over the eyes. Imagine if somebody was vaccinated against smallpox, measles, or tuberculosis, and they still got infected with the disease, became sick, and potentially infected other people…then would it be fair to say that the vaccine did not work?

Go to a doctor in 2018 and say '*I've a bad cough and fever*! 'And the reply would be, '*it's a winter bug stop wasting our time.* Not a mention by politicians or the media that there are tried and tested alternatives which are less risky than taking an experimental shot. Prophylactics like hydroxychloroquine (HCQ) are widely endorsed by doctors who have had success with the treatment. According to Pierre Kory, MD ICU Pulmonologist—Ivermectin appears to stop the illness or its symptoms falsely labeled as COVID-19. *This indirectly show that illness is not virus based, but could be practice for which intersection is standard treatment.*

The Zelenko Protocol (low doses of zinc sulfate, hydroxychloroquine (HCQ), and azithromycin) is one of the most prominent. While another alternative, Quercetin is a good, cheap ($2) freely available (from Amazon) known COVID treatment. It has similar effectiveness to Merck's new ($700) pill (just below it). But these cheap generic (not profitable) remedies are vilified, ignored, or banned.

Worse yet, in December last year, the second-largest hydroxychloroquine factory in the world inexplicably burned down.[58]

On December 14, 2021, UK Prime Minister decided one reported Omicron death (though sad) was worth ordering all doctors to delay all other medical appointments and focus on booster jabs—with the 'Moronic' variant being widely reported as mild. This insane step must be factored against 50,000 pending preventable deaths from undiagnosed cancers.

That number will be bigger as more support medics are diverted to the booster campaign. But again, will anyone in government or Big Pharma explain what is in the booster jab that protects against Omicron that wasn't in the first two jabs?

None of this is about preventing people from dying, it's about preventing people from living.

Those who have been paying attention will recall that Dr Robert Malone, the man who co-invented mRNA technology, referred to Covid-19 vaccines as Experimental Gene Therapy. Moreover, Malone cautioned against mass jabbing of the general population with mRNA vaccines.

---

[58] https://taiwanenglishnews.com/pharmaceutical-factory-on-fire-after-explosion-2-injured/

For his trouble, the New England Journal of Medicine retaliated by blocking Malone's IP address. What does that say about *follow the science* when you permanently block the guy who helped to invent it?

Malone, like many whistleblowers, is fighting an uphill battle against the evil triumvirate. Big business plays the game of bending the rules and getting away with what it can. Lost on many is that these vaccines are useless simply because the average age of death with COVID is over 80. It's like expecting them to make people immortal.

Science is dependent on grants and so increasingly gives the answers that sponsors require.

Government is in the business of staying in power (unless, of course, you can either cancel the next election, or make so much money in the interval that you don't care if you're voted out).

Given their disparate aims, they don't make for a happy combination when in any sort of suspicious agreement. At least, now we understand why the CEO of Pfizer and the President of Bayer called this a gene therapy, not a vaccine. Prudence dictates that with this modern technology their use should be both limited and targeted to the exceptionally vulnerable.

Conspiracy theory being proven fact as each month passes. We tinfoil-hat-wearers warned that we were witnessing a dangerous gene-altering development, the RNA vaccine is a genetic modification injection. We give a hearty thumbs up to all the unvaccinated. You made a wise choice.

Especially concerning should be clamor from the drug companies that booster jabs are necessary on a regular six-month basis. Your half-yearly shot opens the door to dependency just to keep your sickly body alive.

Keep in mind that COVID-19 used to be called Novel Coronavirus-Infected Pneumonia.

"Novel coronavirus" means the variant of the SARS virus called SARS-CoV-2. Pneumonia means pneumonia.

As a society, we are losing our grasp of the differentiation because people no longer say they have got a cold, they say they've got COVID. It doesn't matter whether they've tested positive for SARS-CoV-2 or not—if they have only got a cold, then they haven't got COVID. (They're unlikely to have flu either unless their symptoms are those of a very bad cold and they're in bed most of the time for at least 5 days.)

As for asymptomatic COVID, there is no such thing. The CDC themselves once said that it is highly unlikely anyone could spread the virus asymptomatically. Then they decided to re-write science.

And what of side effects in years to come from the jabs? Lifetime dependency on pharmaceuticals?

Everybody has untold DNA strands that are the result of past generations infected by pathogens and this has been going on for as long as there have been mammals, humans included. An argument may be made that invasive organisms (parasites) may be beneficial and some neutral and some harmful.

The point being that our immune systems specialise in dealing with those invaders and creating antibodies to defend us.

British data is also forcing the sane among us to question the wisdom of this mass injection campaign.

From a Technical Briefing,[59] Public Health England data showed that the COVID death toll is higher among the fully vaccinated compared to the unvaccinated.

Between February 1, 2021, and September 12, 2021, of 157,400 fully vaccinated patients (26.52% of total cases) were diagnosed with the virus/variant. Among the unvaccinated, there were 257,357 such cases (43.36% of total).

While infections were far more prevalent among the unvaccinated, these patients also had better outcomes. In all, 63.5%

---

59 https://assets.publishing.service.gov.uk/government/uploads/system/uploads/attachment_data/file/1018547/Technical_Briefing_23_21_09_16.pdf

of those who died (allegedly) from COVID-19 within 28 days of a positive test were fully vaccinated (1,613 compared to 722 in the unvaccinated group).

Pharmaceutical companies understand that the consequences are that our laboratory modified DNA will not be able to repair itself and that cancerous tumors will be the result. Moreover, a permanently genetically compromised immune system combined with a drug dependency ensures we have become completely enslaved. We no longer will evolve in harmony with our environment but become manufactured *in vitro*; transhumanism.

While it may be argued that it is not the dreaded Spike Proteins that are DIRECTLY causing the likely increase in all cancers, it is accepted that they are *impeding*—delaying or preventing—the DNA repair proteins BRCA1 (especially important) and 53BP1 (first identified as a DNA damage checkpoint protein.[60]

It is also widely accepted that our body is constantly, and continuously, repairing itself at a cellular level because of damage (genetic mutation) cause by external sources (Radiation—Cosmic, Solar, Artificial—Environmental Toxins, Pathogens, etc.) and internal sources (random mutations, etc.).

As an analogy, consider the damaged cell as a badly injured passenger in a car accident; the DNA Repair Proteins are the ambulance with a Doctor, EMT, etc. inside of it. The Spike Proteins are a traffic jam preventing the ambulance from reaching the crash site and blocking the medical help required to save the injured car passenger.

As such, without the DNA Repair Proteins that function naturally thanks to our immune system, we are not optimal to correct the abnormalities and various forms of cancer will form. Pharma companies that specialize in cancer medications will make huge additional profits. What a 'win-win' for the shareholders!

This more realistic characterisation of the jab reveals that it

[60] https://pubmed.ncbi.nlm.nih.gov/24469398/

was never truly a vaccine and explains why the UN WHO modified the definition of a vaccine TWICE so they could call it a vaccine, as did the CDC in the United States (see below).

Suspicions that this was intended as a gene-altering serum were not helped by the secrecy involved. Pfizer prevailed upon the FDA to have their vaccine trials sealed by court order, but an honest judge unsealed 300,000 pages of documents to reveal harrowing details.

Not a mention of any of it by the legacy press who are fully onboard with the apparent mass genocide. The New York Times felt compelled to issue a massive correction after overstating COVID hospitalizations among children by more than 800,000! The number of U.S. children hospitalized with COVID was 63,000 from Aug 2020 to Oct 2021, not 900,000 as the NYT reported. The article also botched actions taken by regulators in Sweden and Denmark and even bungled the timing of a critical FDA meeting.

Children are over one hundred times more likely to die from these experimental injections than COVID-19. Injected athletes, around the world, are collapsing playing their sport, clutching their chests. Notwithstanding that reporting systems are limited and passive, millions of adverse effects have been recorded, which include death, paralysis, blood clots, strokes, myocarditis, pericarditis, heart attacks, spontaneous miscarriage, chronic fatigue and extreme depression.[61] [62] [63] [64]

This mass of shocking evidence is being ignored by professional 'journalists' in the media. The purpose is clear: to prevent access to information that may dissuade the vaccine-hesitant from taking the shots. We may fairly assert this based on such evidence that YouTube expanded its ban on so-called

---

[61] coronavirus-yellowcard.mhra.gov.uk

[62] vaers.hhs.gov

[63] ema.europa.eu/en/human-regulatory/researchdevelopment/pharmacovigilance/eudravigilance

[64] vigiaccess.org (search covid-19 vaccine)

misinformation to include removing any video that may cause vaccine hesitancy, regardless of whether or not the content in the video is factual.

This includes the removal from the platform of some CDC meetings, as demonstrated by the recent deletion of the ICAN's upload of a February 2018 ACIP meeting where they unanimously approved a Hepatitis B vaccine despite being "concerned" over "that myocardial infarction signal." The trial of this vaccine had 14 heart attacks in the test group and one in the control.

Despite the mainstream media's best efforts to hide the facts the word is gradually spreading that all is not well. An AI-powered US Department of Defense program named *Project Salus*, run in cooperation with the Joint Artificial Intelligence Center (JAIC), analysed data on 5.6 million Medicare beneficiaries aged sixty-five or older and concludes that "the vast majority of covid hospitalizations are occurring among fully-vaccinated individuals" and "outcomes among the fully vaccinated are growing worse with each passing week."

The Health Ministry of Israel (a nation mostly double-jabbed) asks the newly vaccinated to "avoid working out" due to an elevated myocarditis risk. Elite athletes, professional sports men and women are collapsing from cardiac complications in the hundreds, but you will not see it on your TV.

Adverse reactions for no apparent benefit compounded by revelations such as the one hundred Royal Navy crew members who were infected with COVID-19 on board the HMS Queen Elizabeth Warcraft despite all crew onboard the ship being fully vaccinated.

These jabs not only do not offer immunity, but they are also actually harming and killing many tens of thousands. In June 2021, Dr Tess Lawrie, co-founder of the World Council for Health and member of the Council's Steering Committee, courageously described the global crisis and called for urgent action:

> *There is now more than enough evidence on the* [UK] *Yellow Card system to declare the COVID-*

> *19 vaccines unsafe for use in humans. Preparation should be made to scale up humanitarian efforts to assist those harmed by the COVID-19 vaccines and to anticipate and ameliorate medium to longer term effects.*

And it's one rule for us and another for them (the elite). Frontline COVID-19 Critical Care Alliance (FLCCC) discovered, via a whistleblower, that between 100-200 US members of Congress were treated their COVID with Ivermectin over the past 15 months.

When independent medical bodies attempt to speak out they are de-platformed and their message quashed before the public can be made aware. PayPal shut down the FLCCC Alliance's PayPal account at the same time Facebook restricts their account for discussing the science around Ivermectin.

Censorship is rampant, as is the authoritarian thievery of our civil rights. Freedoms stolen in a drip-drip fashion so as not to make it too obvious that we are being enslaved by fear of a disease with a 99.9% survival rate.

Two months before his Twitter account was suspended, Malone wrote a rather prophetic Twitter post:

> *I am going to speak bluntly. Physicians who speak out are being actively hunted via medical boards and the press. They are trying to delegitimize us and pick us off one by one. This is not a conspiracy theory but a fact.*[65]

In a real pandemic, hospitals desperately hire as many doctors and nurses as they can. They certainly don't fire them. At the time of

---

65

https://twitter.com/MaajidNawaz/status/1476532434602795008/photo/1

writing (December, 2021) in the UK there are currently 32,190 vacancies in the NHS, and the British government is threatening termination of employment of 110,000 unvaxxed staff. Is this what you do in a real pandemic?

In a true pandemic, you utilize every able-bodied healthcare professional available. You release any and all available treatments. Because the only thing that matters is saving lives. But that's not what's happening here. There is a sinister motive behind all of this.

When we question the official narrative and choose not to accede to the insanity we are called anti-vaxxers (we're really just pro medical choice). We are not extremists, just a moderate minority and a convenient scapegoat. People who dislike the idea of a future dictated by large corporations and very rich people are not anti-vaxxers. They are people who dislike the idea of a future dictated by large corporations and very rich people.

We are not all in it together because this is a pandemic of people not taking responsibility for their own lives. The moment you surrender control of your body in order to participate in society you are no longer a free citizen.

We are being divided so we can be controlled.

Think about that.

Chapter 13—Relevant Physiological and Biological Concepts

IT IS COMMON knowledge that the body requires carbohydrates, fats, proteins, oxygen, water, and micronutrients such as minerals and vitamins for its maintenance and growth. The question is, how does the body perform this function with the mentioned ingredients or nutrients?

These components are hardly ever administered in their native or pure forms but as food, such as bread, milk, meat, etc. The body breaks them down into molecules (amino acids, sugars, fatty acids), absorbs them, and then rebuilds the desired fat and protein molecules for numerous uses and functions.

It is generally not recognized that the body does not use the food and its components, particularly the large molecules, but breaks them down and rebuilds its compounds for specific needs and functions.

For example:

Larger molecules (carbohydrates) such as starches are broken down in the digestive system, reducing starch molecules to single glucose units. Once these carbohydrates are fully digested, the cell linings of the small intestine absorb them and transport them to the bloodstream.

Fats: The digestive system breaks down dietary fats into fatty acids. The resultant fatty acids get absorbed and passed along to your circulatory system.

Proteins: The small intestine cannot absorb a molecule as large as protein. The gastrointestinal tract, therefore, breaks it down to amino acids.

The cardiovascular system is the organ that plays a critical role in supplying/distributing the necessary ingredients (food, water, air, etc.) to the body. The primary ingredient entry point is the gastrointestinal (GI) tract through the oral cavity. However, other minor entry points such as muscle, rectum, vagina, etc., may also be used for specific purposes such as administering medicines.

Like in chemistry, where the simplest or smallest functional unit is an atom, the body's basic functional unit is a cell. All living structures of human anatomy contain cells, and almost all functions of human physiology happen in cells or are initiated by cells.

A cell has three main parts: the cell membrane surface or boundary, the nucleus, and the cytoplasm (soup) within the cell boundary. The cell membrane surrounds the cell and controls the substances that go into and out of the cell.

The nucleus structure inside the cell containing the nucleolus and most of the cell's DNA. It is also where most RNA is made.

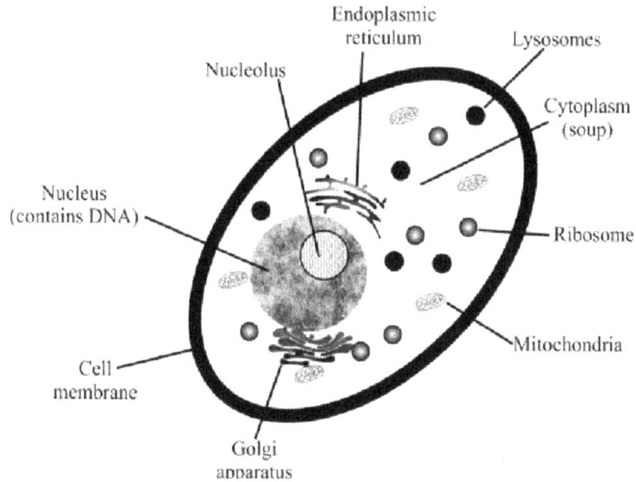

Figure 36
Schematic Representation of a Cell and its Composition

The cytoplasm is the fluid (soup) inside the cell. In addition, it contains other tiny cell parts that have specific functions, including the mitochondria and the endoplasmic reticulum. Most chemical reactions occur in the cytoplasm, where most proteins are made. Mitochondria are complex organelles that convert energy from food into a form that the cell can use.

Basically, a cell is a fully self-supported chemical manufacturing facility, with different "departments" assigned to conduct specific jobs like any worldly manufacturing plant. The basic biology and physiology textbook can provide detailed descriptions of cells' structure, composition, and functions. In humans, as in all organisms, cells perform all life functions.

Pharmacokinetics, abbreviated as PK, describes the time course of drug absorption, distribution, metabolism, and excretion. The substances of interest may include pharmaceuticals, pesticides, food additives, cosmetics, etc. Pharmacokinetic studies attempt to analyze chemical metabolism and discover the fate of a chemical from the moment it is administered up to the point at which it is eliminated from the body.

From the medical perspective, the primary goal of pharmacokinetics includes studying the efficacy and toxicity aspects of a patient's drug therapy. In addition, developing strong correlations between drug concentrations and their pharmacologic responses enables administering the appropriate amount of drug (dose) to patients.

A drug's effect is often related to its concentration at the site of action. However, direct measurement of drug concentrations at the sites of action is not practical. Therefore drug concentration in blood levels is considered a surrogate for the drug concentration at the site of action.

Anyone involved in therapeutic drugs must understand the following two that explain the link of chemistry with biology and physiology.

Pharmacology is a discipline that studies drug action, where a drug exerts a chemical and/or physiological effect on the cell, tissue, organ, or organism. More specifically, it studies the interactions between a living organism and chemicals that affect normal or abnormal biochemical function.

Pharmaceutics is the subject that deals with dosage forms (tablets, capsules, injections, ointments, etc.) used by patients for establishing the delivery and disposition of drugs in the body.

The following describes an overview of these topics related to drug development, manufacturing, and evaluation of their safety and efficacy profiles.

Before considering the evaluation aspects of medicines and medicinal products, it is essential to clearly understand the difference between them. Unfortunately, the public, including most professionals and experts, are unaware of this difference and make scientifically invalid claims. Therefore, understanding the true nature of medicines is imperative, and people should have some basic understanding of the underlying science of the subject.

Medicines, in general, are pure, isolated, and fully characterized chemical compounds. Below are some commonly prescribed medicines with their names and chemical structures.

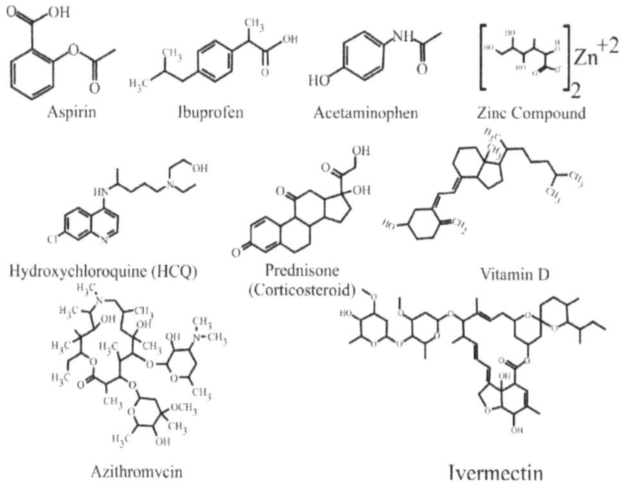

Aspirin Ibuprofen Acetaminophen Zinc Compound

Hydroxychloroquine (HCQ) Prednisone (Corticosteroid) Vitamin D

Azithromycin Ivermectin

Figure 37
Chemical Names and Structures of some Frequently Refereed Medicines during the Pandemic

These chemicals (medicines) are generally available from many chemical suppliers meeting quality standards, often exceeding regulatory authorities' requirements. The critical thing to note here is that although these chemicals are considered and regulated by health authorities worldwide as medicine, it is not part of the medical discipline. Instead, it is purely chemical development, characterization, and manufacturing aspects.

It is like the manufacturer of plastic bottles (product) that receives the raw plastic material, often in powder or pellet form, with predefined and desired specifications. The plastic bottle manufacturers mostly have no or limited expertise in the manufacturing process of the raw plastics and setting its specifications.

However, a medicinal product is a composite of multiple components, medicine being one. In this respect, medicine is an active ingredient while other inactive ingredients or excipients.

For example, candy is not sugar but contains sugar as one of its ingredients. Similarly, flour is not bread or cake but is one of the (major) components but includes oil, fruits, and nuts and processing/baking.

In addition, medicinal products undergo different manufacturing, testing, and regulatory approval protocols than active ingredients (medicines).

The public or patients buy medicinal products (like plastic bottles) and want assurance that the products are of the expected performance, i.e., safety and efficacy. Therefore, considering their mandates, the regulatory authorities are supposed to provide such assurance. The question is, do they? The answer is no. With all their claims of their medical expertise, they do not and cannot, as explained below.

Pharmacology and pharmaceutics are considered separate disciplines (as described above). However, they overlap so much that it is impossible to describe them separately.

However, pharmacology generally deals with pure active ingredients (drugs), while pharmaceutics deals with drug products.

They both use the same techniques and remain forms of chemical testing, i.e., testing medicines (chemicals), and must follow the principles and practices of testing described earlier.

Medicines and Medicinal Products Testing

Such studies are also known as pre-clinical (animal) and clinical (human) studies.

These studies are conducted in whole animals. These studies are conducted to evaluate physiological impacts (cell and tissue damage) and the fate of the medicine in the body. They are assessed by studying the absorption, distribution, metabolism, and excretion characteristics of the drugs in the body, along with toxicological/safety profiles.

Clinical testing/trials or in vivo testing is a critical and indispensable part of developing safe and effective medicines. Therefore, ignoring such studies, as with COVID-19 vaccine development (explained later), may be considered a negligence case.

Pharmacokinetic studies are conducted mainly to evaluate the movement of the medicines and their metabolites through the body. The responses (pharmacodynamics) depend on the drug concentrations in blood (often monitored in plasma), which change with time. Therefore, it is imperative to know the drug levels to maintain desired therapeutic concentrations. These studies are based on analytical chemistry testing and are undertaken extensively on animals and humans. Interestingly, pharmacokinetic studies are surprisingly simple in concept and practice.

Clinical trials (testing in humans) are often presented as a highly sophisticated, specialized, and modern approach to developing and manufacturing medicines. However they may be described as typical and standard analytical testing.

Clinical trials are comparative evaluations of an entity (usually a chemical-based medicine irrespective of its size, small or large molecules, or source chemical, plant or biological) against a blank or control (commonly known as placebo). Hence, it should fall under the category of the science of testing. A clinical trial is called clinical because instead of testing (trials) in animals or non-animal laboratory-based, they employ human subjects. The subjects (humans) are divided into two groups, one for the treatment and the other without, to see if the medicine provides the expected outcome without producing unacceptable adverse effects.

One can do a simple clinical trial at home to understand the principle of clinical trials. For example, a study (a clinical trial) can easily be conducted at home to observe weight gain with sugar consumption. To run such a clinical trial, one would need some healthy volunteers, a supply of sugar (e.g., chocolate-loaded ice cream), and a weighing scale to monitor the weight. All one has to do is divide the volunteers into two groups and feed one of the groups a bowl of chocolate-loaded ice cream daily for a month. Monitor the weights at a predetermined schedule and establish the weight differences. Voila, one has done a clinical trial.

Clinical trials are conducted in a "clinical" setting with a control study design and laboratory tests to monitor the effects with or without medicines. In addition, however, they require identified medicine or illness and a quantifiable outcome, using a monitor/test.

In vitro testing uses cell lines from different sources such as animals and humans, even microbial or plant-based. Therefore, proper testing requires well-defined test compounds and valid tests with clearly measurable endpoints that may reflect in vivo (chemical or physiological) endpoints.

In vitro tests are without animal or human use. They may be visualized by considering a simple experiment with two solutions in two separate test tubes—one containing silver nitrate ($AgNO_3$) solution and the other common salt (sodium chloride, NaCl). One would immediately observe that the solution becomes cloudy after mixing these solutions in one of the tubes. This cloudiness represents the result of a chemical reaction, conversion to silver chloride (AgCl) and sodium nitrate ($NaNO_3$). There would not be any common salt in the solution anymore. The cloudy white material is formed by AgCl, which is insoluble in water and comes out of water. This is an example of a chemical reaction with visually observable output. Most chemical reactions are not simple to conduct or observe, and therefore some sort of testing is required to follow the reaction.

Similarly, if one puts some cellular or microbe specimen in water with chemicals of interest (substrates) and some nutrients for the cell to live on, these cells will cause chemical reactions to occur. Unlike the example of the above chemical reaction of silver nitrate and sodium chloride, which occurred instantaneously at room temperature, biological reactions, in particular in humans, mainly occur at 37°C (normal body temperature). It is important to note that this cellular reaction may occur within the cells or external to them with chemicals (enzymes) produced by the cells and transferred into the test tubes or flasks. The reaction content would appear in the tube by transferring molecules in and out of the cells. These chemical reactions within cells occur with the help of enzymes.

A simple and daily life example of an enzymatic reaction would be the conversion of starch into sugar. Assume, for example, that a person has just eaten a piece of bread. An important nutrient in that bread is starch, a complex carbohydrate. As soon as the bread enters a person's mouth, digestion occurs. Enzymes in the mouth break down starch molecules and convert them into smaller molecules of simpler substances: sugars. This process can be observed easily by holding a piece of bread in the mouth and observing a sweet taste—the taste of the sugar formed from the breakdown of starch.

Such in vitro tests are often criticized for their lack of relevance to human physiology, which is true. However, such cellular metabolic studies are critically important to form an educated opinion on what one should or could expect in humans. For all practical purposes, such testing should not be considered with a higher degree of predictability of human output but for putting pieces together. Thus, one observes metabolic outcomes and the toxic effects on cell morphology and or deaths.

Outside the pharmaceutical industry, adequate reference standards or endpoints based upon human data are rare. Consequently, human cell-based tests are often developed against animal origin.

Whether in vitro tests are based on primary cells, cancer-derived cell lines, or stem cells, it is essential to have in vitro systems that often adequately mimic key events of the in vivo actions triggered in humans upon exposure to a medicine.

The cell functionality in vivo (humans and animals) is driven by the microenvironment surrounding the cell and cell-cell interactions, usually lacking in vitro. Therefore, the development of in vivo-like in vitro tests requires understanding the impact of such a microenvironment.

A more in-depth understanding of the relationship between toxicity and biological pathways makes it possible to prevent or at least reduce animal and human testing.

The previous description provides the fundamental science part of medicines. The part explained with relevant details highlights that medications, including viruses, DNA/RNA, and proteins, belong to chemical science, at least for their development, isolation, and characterization.

Based on the described fundamental scientific principles and accepted practices, it should be clear that the virus, RNA, and spike protein have never been isolated. Furthermore, no scientifically valid test method can be developed for COVID-19-related substances without their isolation.

Unfortunately, medical experts and practitioners ignored the actual underlying science. Instead, they developed a version of pseudoscience to implement their rules and regulations. Unfortunately, this pseudoscience created imaginary viruses, invalid tests, and non-existing pandemics, followed by false and fraudulent treatments, particularly vaccines, as described in the following sections.

Chapter 14—The Regulatory Authorities, The Source and Promoters of Pseudoscience

SO FAR, WHAT has been described is the scientific and physiological aspects of the medicines. Appropriate detail has been provided to demonstrate that drug and drug product development, manufacturing, and assessments are chemistry-based. It is not intended to teach chemistry here in depth but to provide a needed understanding of the science behind medicines.

An important point to note here is that the word chemistry is used to reflect science. Subjects named by adding a suffix "science" should not be considered scientific disciplines but occupations (trades) that use well-established science principles, such as chemistry.

Political science, computer science, data science, pharmaceutical science, medical science, health (epidemiological) science, clinical science, and regulatory science are examples of such subjects, which certainly create significant confusion in following the science. Conversely, humans' medicines (development, manufacturing, and evaluation) should be considered a sub-discipline of chemistry, i.e., studying chemical compounds and their interactions (chemical reactions) in the body.

It is commonly assumed that consumers or patients lack an understanding of the complex science (chemistry) vocabulary. Therefore, someone on their behalf should work to protect their interest. In modern-day societies, this role is fulfilled by governments through organizations such as Health Canada, the United States Food and Drug Administration (US FDA), European Medicines Agency (EMA), etc., along with some international organizations such as World Health Organization (WHO). Practically all countries have such bodies. The main goal is to watch the medicines' safety, efficacy, and quality on the public and patients' behalf.

The safety, efficacy, and quality aspects are regulated through the Acts and Regulations of parliaments and managed by bureaucracy. The administration implements the Acts and Regulations in essence and spirit through guidelines or guidance (standards or requirements). There are an infinite number of such guidance documents worldwide from regulatory authorities.[66] [67] [68]

In addition, these documents constantly change and proliferate as per the needs of the bureaucracy. In general, these may be divided into two categories to manage the drug approval process (1) safety and efficacy; (2) manufacturing (commonly known as Chemistry and Manufacturing Controls or CMC). In reality, the safety and efficacy part is mainly done at the beginning of the medicine development. As explained earlier, medicine development is dominated by the isolation and characterization of chemical compounds. However, CMC is the permanent and continuous part of overseeing the medicines' manufacturing and availability—note *manufacturing*, not the safety and efficacy.

It may be helpful here to explain what medicine development means. It means linking an entity (mostly a chemical substance) to some beneficial physiological outcome (endpoint) such as fever, blood pressure reduction, and addressing higher sugar levels.

For example, suppose someone observes that a freely available chemical substance appears to reduce the blood level in humans. This would be considered a lead for a further investigation called a

---

[66] https://www.fda.gov/regulatory-information/search-fda-guidance-documents/clinical-trials-guidance-documents

[67] https://www.fda.gov/animal-veterinary/guidance-industry/chemistry-manufacturing-and-controls-cmc-guidances-industry-gfis

[68] https://www.fda.gov/media/71023/download, https://www.fda.gov/files/drugs/published/Guidance-for-Industry-Bioavailability-and-Bioequivalence-Studies-for-Orally-Administered-Drug-Products---General-Considerations.PDF

clinical trial in which the lead substance is administered to patients against the blank (placebo). The outcome (beneficial effect) would be monitored by measuring the blood pressure (tester).

The test or tester is usually developed and validated by third parties. If the tested entity is found effective in reducing blood pressure without observable or unacceptable adverse effects, then it is considered that a medicine has been developed. Note that the substance tested is not new (it does not have to be) and may be freely and cheaply available, so nothing is created or developed, but the tested entity has been linked to a beneficial physiological outcome. This is how our ancestors developed medicines, and this is how we are developing them today, with extremely rare exceptions. However, unfortunately, the public perceives that drugs are being developed based on the most modern and advanced scientific principles and practices.

In reality, that is not the case. The public assumes that "pharmaceutical scientists" know and apply the body's chemistry, physiology, and deviation during illness for developing medicines. The chemical entity takes part in that reaction and brings it back to normal conditions. However, that is not the case. It is still based on observations and survey.

Certainly, pictures or chemical structures (proteins, DNA/RNA, antibodies) are shown to give the impression that science has been followed. However, such images are often without substance and are like graffiti on the walls. A recent example of this kind of science is showing pictures of the virus and its RNA and spike-proteins. No one has isolated the virus, let alone its RNA and proteins. How could the science be done—with these entities? It can't be. Moreover, testing claims have been made for all these imaginary images.

Also, most of the practices in the regulatory area relate to product development, manufacturing, and marketing like any other industry of manufacturing chemical ingredients and their composites/products. Interestingly, the medical or regulatory authorities develop and impose their versions of manufacturing guidelines. Manufacturing guidelines developed by medical authorities and experts! Under the heading of FDA's Current Good Manufacturing Practices (CGMP), Guidance for Industry Quality Systems Approach to Pharmaceutical CGMP Regulations.[69]

The very first line of the guideline document states:

> *This guidance is intended to help manufacturers implementing modern quality systems and risk management approaches to meet the requirements of the Agency's current good manufacturing practice (CGMP) regulations (21 CFR parts 210 and 211).*

This is about establishing the quality (management) system for manufacturing—the word quality is mentioned 338 times in the 32-page document. So, what do the guidelines, in reality, mean? These are protocols (set of instructions, more like rituals) to be followed by the manufacturers to achieve quality products the authorities would approve.

All are under the guidance and control of medical and pharmaceutical professionals and experts to monitor medicines' safety, efficacy, and quality. However, it may be shocking to know that the safety and efficacy of medicinal products are hardly ever monitored during the commercial production of the products which consumers/patients buy and use. Instead, only the quality, which serves as a surrogate for the safety and efficacy of developed drugs, is clearly described at one of the FDA sites.

> *Pharmaceutical quality is the foundation that allows patients and consumers to have confidence in the safety and effectiveness of their*

[69] https://www.fda.gov/media/71023/download

medications.[70]

As noted above, there are an infinite number of regulatory guidelines and standards that the medicinal/pharmaceutical industry has to follow. It is neither possible nor needed to describe or explain drug and drug product evaluation and approval systems in depth here. It is so horrendously lengthy, complex, and confusing that even working professionals often do not understand the requirements. That is why it is usually recommended that manufacturers of medicines hold meetings with the FDA or other countries' authorities to seek clarification and specific expectations. It means that evaluation and approval are based on subjective interpretation of the guidelines, and the country's acts and regulations, by extension, can be molded or deviated as per need.

Hence, collectively explaining regulatory guidelines and standards would be more effective and productive. Moreover, it will clearly demonstrate that the current regulatory practices and standards are neither scientific nor assure product safety, efficacy, and quality but ritualist practices.

The US is recognized as the leading authority in medicines, product development, and assessments. Therefore, others follow FDA's approach and develop their guidelines and standards accordingly. Often it is like managing a script, which may differ in shape, size, and font, but the content remains the same. Therefore, FDA approaches are used here to describe the requirement and standards.

The main problem is that not all are based on science (chemistry) but are written with little understanding and knowledge of chemistry/science.

The center known to evaluate new drugs before they can be sold in the US is the Center for Drugs Evaluation and Research (CDER). CDER's job is to prevent quackery and provide physicians

---

[70] https://www.fda.gov/drugs/development-approval-process-drugs/pharmaceutical-quality-resources

and patients with the information to use medicines wisely. In addition, the center ensures that both brand-name and generic drugs work correctly and that their health benefits outweigh their known risks. This and sister organizations worldwide are self-regulating and, sadly, mostly independent of any third-party audits for all practical purposes.

The following discussion provides an overview of regulatory authorities' requirements to show that drugs and products are safe and efficacious for public use, hence marketing approval.

First, from the pharmaceutical products evaluation perspective, let's see the description by the FDA guidelines under Abbreviated New Drug Application (ANDA) states:

> *An abbreviated new drug application (ANDA) contains data which is submitted to FDA for the review and potential approval of a generic drug product.*[71]

One of the guidelines, perhaps the most important and extensively followed in the industry, is on ANDA (Abbreviated New Drug Application). However, the applications are not about new drugs but new products or different products of an old drug. The guideline provides the basis for comparing different drug products, i.e., formulation and manufacturing evaluation, not the drug.

However, the medical and pharmaceutical authorities consider them as new drugs in ignorance.

Evaluating a new drug requires a clinical trial for safety and efficacy, but these new products do not. Furthermore, they require only equivalency of products, i.e., equivalencies of the formulations, not drugs. Therefore, the authorities give a false impression that they approve and provide new and cheaper pharmaceuticals.

---

[71] https://www.fda.gov/drugs/types-applications/abbreviated-new-drug-application-anda

So, if the products are not based on clinical trials, how are they assessed for their safety and efficacy? Instead, they are based on their quality assessment, i.e., if the quality is good, then products are good, i.e., safe and efficacious.

So, how does the quality of the products is established? It is not! There is no acceptable and official definition of product quality anywhere, including from regulatory authorities.

However, just like in the case of manufacturing, authorities set protocols (rituals) for establishing the quality of the products, which are irrelevant and arbitrary but based on peers-reviews and advice from expert committees.

The three protocols are commonly followed in this regard:

(1) mainly analysis of active ingredients,
(2) bioavailability/bioequivalence assessment, and…
(3) drug release or dissolution testing.

Understanding these protocols is of critical importance, with the knowledge of chemistry and physiology aspects described earlier, to indicate that following these protocols is neither based on science nor establishes the quality of pharmaceutical products. Furthermore, it will also help understand how the "scientific" (ritualistic) practices at the FDA and CDC resulted in the fake COVID-19 virus and its vaccine debacle.

The following tests and acceptance criteria are generally considered applicable to drug products:
Description: A qualitative description of the dosage form should be provided (e.g., size, shape, and color).

Identification: Testing should establish the identity of the drug substance(s) structure using techniques such as HPLC and GC against reference standards.

Assay: To demonstrate the strength (content) that should be included.

Impurities: Organic and inorganic impurities (degradation products) and residual solvents are included in this category and should be monitored and reported.

Physicochemical specifications should well characterize primary reference standards. Working reference standards also need to be described, and that description should include the preparation, characterization, and specifications of the working reference standard.

> *Reference standards are highly purified and well-characterized compounds that are essential for validating the quantitative and qualitative accuracy of analytical testing methods.*
>
> *Thus, reference standard qualification, development, characterization, and management is critical for companies working to deliver safe and efficacious therapeutics that can withstand regulatory scrutiny.*[72]

> *In some cases, well-characterized, high-purity (>99.5%) compendial reference standards—often called primary reference standards—are available for purchase from pharmacopeias, such as the USP or other national formularies.*[73]

The reference standards noted above are available from the USP (United States Pharmacopeia). However, this is incorrect because the USP does not provide any product standard. It only provides standards of pure chemical compounds labeled as medicinal standards with a warning that these are not for human or

---

[72] https://www.fda.gov/media/87801/download

[73] https://www.avomeen.com/scientific-applications/reference-standard-characterization-and-qualification/

therapeutic use.

> *USP Reference Standards (RS) are not for use in humans or animals as drugs or medical devices. They are intended only for use in analytical or laboratory applications generally as specified in USP compendia.*[74]

It is, therefore, very important to note that any testing of pharmaceutical products is conducted *without* a reference standard, i.e., only following the protocol (compliance) is the requirement, not monitoring or establishing the quality of the product against any reference standard. Scientifically speaking, this makes all the pharmaceutical products' testing and quality claims irrelevant and invalid and must be withdrawn.

Regulatory authorities, in particular for developed countries, require two sets of testing to establish the presumed quality of pharmaceutical products, particularly oral dosage forms.

Bioavailability/bioequivalence assessments: The testing is based on the principles of pharmacokinetics, as described earlier. A protocol is provided as an example in this respect.

A bioavailability study is usually conducted in healthy human adults, involving 12-30 subjects. Variability in results is minimized by conducting studies on healthy human volunteers. They should meet specific characteristics such as having an age between 18-50 years, falling within the normal weight range, and being free from disease or using any medication. The institutional ethics committee must approve the study protocol.

Blood samples (10 to 15 in number) are withdrawn from the volunteers following the dose administration. The sampling times' choice should be such that C_{max} (highest concentration observed in

[74] https://www.usp.org/sites/default/files/usp/document/our-work/reference-standards/qa-policy-statement-usp-rs-uses-and-applications.pdf

the blood) and T_{max} (when highest concentration is observed) could be reasonably established. In addition, the sampling duration should be such that most (at least 80%) of the drug can be accounted for based on the area under the blood-concentration-time profile (AUC).

The most commonly used testing methods are chromatographic such as High Pressure/Performance Liquid Chromatography (HPLC) or Gas Chromatography (GC). Pharmacokinetic parameters from the plasma drug release profiles are determined for individual volunteers. The average values of these parameters would reflect the bioavailability of the product. A typical protocol for conducting a bioavailability study is described in a Health Canada paper.[75]

A bioavailability study compares the bioavailabilities of reference vs. test products. A test product may be changed formulations and/or manufacturing attributes or a completely different product such as a generic. The generic industry brings products to the market by conducting bioequivalence studies between an innovator's (reference) and generic (test) products.

The study has to be conducted following a recommended statistical approach and design—usually—in a crossover design.

Once the comparative bioavailability study's experimental part is completed, the respective pharmacokinetic parameters are derived and compared. The products are declared bioequivalent (interchangeable) if they meet the recommended regulatory specifications for the parameters. The parameters and required specifications may vary from country to country. However, the most common standard followed by the pharmaceutical industry is that of the US FDA. In this case, a 90% confidence interval of the ratios of the log-transformed values of parameters (C_{max} and AUC)

---

[75] https://www.canada.ca/en/health-canada/services/drugs-health-products/drug-products/applications-submissions/guidance-documents/bioavailability-bioequivalence/conduct-analysis-comparative.html

should fall within 80-125.

These bioavailability and bioequivalence studies are considered gold standards to establish drug product quality—note product, not a drug.

Such studies are also conducted for other types of comparison, such as drug-drug interaction, food-drug interaction, etc., to evaluate if concurrent administration affects the blood levels of a drug.

The above-described bioavailability and bioequivalence studies are done before commercial production of the products. During commercial production, such studies are seldom conducted. Therefore, product release characteristics or quality are generally established based on in vitro assessments.

In vitro (or drug release or dissolution) testing: Such studies have generally required USP methods or protocols. For example, the most widely recommended dissolution test methods are based on the USP Basket and Paddle apparatuses.

Dissolution testing is normally carried out under physiological conditions, such as an aqueous medium maintained at 37 C with softer stirring, allowing appropriate interpretation of dissolution data concerning in vivo performance of the product.

Chapter 15—Further Explaining the Pseudoscience at the FDA

IT SHOULD BE clear from the discussion provided that medicines are chemicals or chemical compounds. The preceding information is based on what is described in medical and regulatory literature, without indicating that the science aspect is, in fact, chemistry—the science of chemical molecules.

Some examples of no or fake science are provided to explain the situation currently causing enormous confusion and errors in the pharmaceutical assessment area.

1. As explained, the body maintains itself through cells' structure and metabolism (chemical reaction) by consuming needed nutrients.

When someone feels ill, it indicates that there may be some issue with the body's normal functioning, i.e., deviation in the natural body chemistry/chemical reactions. Therefore, medicines (chemicals) are prescribed to normalize the body's function. However, the issue is that most, if not all, chemical drugs are unnatural (outside the normal body's needs or requirements). Therefore, they are bound to produce strange and damaging effects (AKA adverse effects). This fact is well known in the medical community.

> *Our prescription drugs are the third leading cause of death after heart disease and cancer in the United States and Europe. Around half of those who die have taken their drugs correctly; the other half die because of errors, such as too high a dose or use of a drug despite contraindications.*[76]

---

[76] https://pubmed.ncbi.nlm.nih.gov/25355584/

This aspect may be explained by considering an ill-functioning machine (e.g., an overheated car) in the middle of nowhere. It was observed that the engine had a crack causing the oil to leak hence the overheating. As the car engine cannot be fixed to actual specifications, a temporary fix is done by tapping on the leak and filling the engine with coconut oil, which was available.

It will certainly make the car work for the time being to bring it to the workshop. Otherwise, the "remedy" will cause severe other problems (adverse effects) because the car is not designed to work with duct tape and coconut oil.

Similarly, the use of chemical medicines such as Tylenol (acetaminophen) and Advil (ibuprofen) are temporary "remedies" to buy time by feeling good. However, these remedies are not the body's natural requirements and are bound to cause other effects (minor or major), particularly with long-term use.

Therefore, one must keep this thought in mind that logically and scientifically, treating people with chemicals unrelated to the body's natural needs by definition becomes a harmful and invalid approach. Furthermore, calling chemicals medicines does not change their characteristics.

Unfortunately, however, medical practice lacks this understanding. Instead, it treats the patients and illnesses with chemicals, often on a long-term basis.

If the treatment has to work scientifically, one must know the body's processes (chemical reactions and imbalances) and fix them accordingly. However, unfortunately, current medical practices do not consider the actual chemistry aspect of the body and its disease. Hence, pharmaceutical treatment could arguably cause more harm than good, at least in the long run.

2. Whenever asked how the products' quality, safety, and efficacy are established and monitored, the usual response is that there is a battery of tests, including clinical trials, available in this respect. However, this is false because no clinical testing is done at the commercial or production levels. None! Product quality, safety,

and efficacy are based on opinions and faith, i.e., compliance with regulatory guidelines, not science. Products are considered safe and efficacious because they are approved (compliant or stamped), not based on their actual performance.

3. Now let us consider how the suggested tests are flawed in establishing the safety and efficacy of the products.

There are two types of tests as described above (1) bioavailability or bioequivalence assessment); (2) in vitro testing or drug dissolution testing.

It is often presumed that bioavailability assessment in humans determines the release characteristics of the medicine in the body.

The bioavailability assessment is based on the assumption that the plasma drug levels are directly linked to the absorption of the drug from its solution form in the GI tract, particularly the small intestine. The higher the absorption (directly related to the drug dissolution/release from the product), the higher the plasma drug levels and vice versa.

It is important to note that the appearance of a drug in plasma from a product is at least a three-step process. First, once a person takes a tablet, capsule, or even a suspension or solution, it goes into the stomach, where the product's disintegration/dissolution starts. It is generally accepted that absorption hardly occurs here.[77]

For the absorption of a drug to happen in the body, the drug has to move into the intestine. The transfer of the drug from the stomach to the intestine is dependent on pushing the stomach content into the intestine (commonly known as stomach motility or emptying effect). The slower the motility/emptying rate, the slower the drug would be available for absorption and vice versa. It is highly unlikely that the entire drug would appear in the intestine simultaneously, especially in different subjects. It usually comes out

[77] Washington, N., Washington, C., and Wilson, C. *Physiological Pharmaceutics: Barriers to Drug Absorption*, 2nd ed. CRC Press, 2001. (p. 82)

in small portions and at random. It is commonly accepted that the stomach emptying time is about three hours but it can vary. Therefore, even if a product is of an immediate release type, it would not appear as a quick-release product because it could take up to three hours for the entire drug/product to appear in the intestine/plasma.[78]

Secondly, once the drug is absorbed from the intestine, it will pass through the liver before appearing in the plasma/blood. The liver metabolizes the drug, often 0 to 60%, depending on its nature and rate of availability to the liver. Therefore, drug dissolution/absorption should not be considered linearly related to the appearance of the drug in plasma as commonly assumed.

To summarize the above in vivo dissolution-absorption-plasma drug levels discussion, it can be stated that it is dependent on three variables: (1) drug release/dissolution from the product, (2) stomach emptying, and (3) liver or hepatic metabolism. The drug dissolution/release is the product characteristic, which is often the least variable of the three—the other two are biological or physiological and are considered to provide higher variabilities.

Indirect evidence of such, i.e., biological/physiological variability, can be provided with a description of a virtual experiment[79] which clearly indicates a high variability of plasma drug levels (~29% RSD) without any contribution from the product or dissolution results. Therefore, commonly observed variability in the bioavailability/bioequivalence studies reflects the combined effect of stomach emptying and metabolism; however, commonly assumed, at authorities, as that of the product.

The shocking outcome of the description indicates that the medical and regulatory authorities' research establishment assumes this variability is coming from the product. But, in reality, it is just

[78] https://onlinelibrary.wiley.com/doi/abs/10.1111/j.1440-1746.2006.04449.x

[79] https://www.drug-dissolution-testing.com/blog/files/disso-to-bio.pdf

the body's expected variability.

In short, bioavailability/bioequivalence, a common regulatory requirement, never determines true drug release characteristics or their variance and, by extension, the safety and efficacy of the products.

It is such a bizarre situation that these irrelevant bioequivalence studies costing 200K+ and dosing potentially toxic chemicals (medicines) to healthy humans without any benefit. In addition, consumers get false assurance that the products are equivalent (safe and efficacious) based on these "scientific" or clinical testing.

Another thing to remember is that the acceptable tolerance is usually set at a 20% difference between test and reference product while expected or natural variability is higher, so bioequivalence studies would often fail, and they do. Unfortunately, rather than looking at the cause of the issue or failure, regulatory authorities often blame the manufacturers for not correctly conducting the study. So, the manufacturers try to match the ingredients and manufacturing processes as close as possible to the reference (innovators') products and repeat the testing until the testing meets the regulatory requirements or standards.

Generic product manufacturing is described with a cartoon version comparing the manufacturing of cupcakes, which is, in reality, the mixing and cooking of commonly available ingredients. The generic product manufacturing would be translated into producing carbon copies of the decades-old products.

Therefore, no change or improvement is allowed.

Most bioequivalence tests are conducted before commercial-scale production at the product development stage. However, the in vitro drug release or dissolution test, a worldwide regulatory requirement is conducted instead of a bioavailability/bioequivalence assessment to establish the quality of the products and, by extension, their safety—and efficacy. However, looking into the test in some detail would reveal that this test is scientifically invalid and irrelevant and can never show the quality of the product, as explained below.

One can do a little experiment at home by having some water in a glass, dropping a tablet of Tylenol, and stirring with a spoon. By doing so, one has done a dissolution test. One can safely drink the water portion of the mixture, which will contain the drug leaving behind the unnecessary residue (inactive substances) in the glass. What occurs in glass occurs in the GI tract. However, in the laboratory environment, the test is conducted in a more standardized manner. Instead of a glass, one uses a round bottom vessel. A spoon is replaced with a T-shape stirrer. Rather than drinking the water, one takes the sample out and measures the content of the drug in the water using any of the testing techniques described in the earlier section, such as (chromatographic).

Once this experiment or study is completed, one claims that the drug product can release the drug as expected and is capable of providing its therapeutic effect. All regulatory agencies require this standard test for all tablet and capsule products worldwide. Hundreds, if not thousands, of test protocols are available in the literature.

So, what could be wrong with this test?

Simply, everything.

The main thing is that this test has never been validated to determine the dissolution of any product. In simple terms, if one is given a blinded sample of a tablet or capsule product, it is impossible to determine the dissolution characteristics of a product. Certainly, one can do the dissolution test as described above, but it cannot provide relevant and accurate drug dissolution results. For this purpose, the test has to be validated using a reference tablet or capsule product with known drug release characteristics. There is none available.

Pharmacopeias, in particular, USP, is considered a world authority in providing pharmaceutical reference standards. However, they do not provide a single reference product with known dissolution characteristics (or quality standards) relevant to human drug efficacy.

Absolutely none!

Therefore, any claim made by the regulatory authority to establish the quality of drug products will remain false and bogus. Hence, no one is monitoring the product quality as claimed.

Every product comes with its test, experimental conditions, and criteria for declaring quality, leading to hundreds and thousands of tests. In simple terms, every box or item comes with its weighing scale to establish the weight of its content. But unfortunately, no one knows the correct weight of any of the items.

On the testing side, the repeatability of the test is so poor that variability of 30% coefficient variance is acceptable for the "high" quality manufacturer product. These drawbacks are thoroughly investigated and published in the scientific literature.[80] However, unfortunately, USP and regulatory authorities do not require reporting coefficient of variation (a standard statistical practice and requirement). Still, they have strange and ritualistic criteria based on individual unit results without any statistical validation.[81]

A few days ago, FDA responded, as a final response, to my (Saeed Qureshi) petition submitted on October 2, 2018, denying the petition. The response acknowledges the flaws in FDA requirements, as highlighted in the petition. However, unfortunately, the petition was denied based on the irrelevant discussion on method development and validation, not on apparatuses/testers validation—the subject of the petition.[82]

The petition stated:

> *...as per cGMP requirements every equipment, including apparatuses/testers, used for drug product manufacturing and evaluation must be validated [e.g. see 6-8] to demonstrate that the equipment or testers are qualified and validated*

---

[80] https://pubmed.ncbi.nlm.nih.gov/9845813/

[81] https://bioanalyticx.com/usp-tolerances-in-terms-of-rsd-or-cv/

[82] https://www.regulations.gov/docket/FDA-2018-P-3742/document

> *for the intended use and purpose. However, the drug dissolution testers have never been validated for their intended use.*

In the response from the FDA, denying the petition after almost 4-years of wait, it is stated that "Method validation and verification encompasses the apparatus used in the method; ***the apparatus is not separately validated.*** *"* [emphasis added] Therefore, confirming that apparatuses/testers are not validated, i.e., FDA (product quality) requirements are based on non-validated apparatus or testers. It infers that all methods developed using such testers and the results obtained have to be non-validated—null and void.

The 7-page FDA response covers the method validation, which is separate from the apparatuses/tester validation and is irrelevant to the requested action to the petition.

So, to conclude, FDA requirements for assessing the quality of products using the drug dissolution technique are based on non-validated apparatuses and testers, not in line with FDA's cGMP requirements. It is a clear example of incompetence and ignorance of the science of testing/testers and its validation at the FDA. An independent third-party audit in this regard is urgently needed.

> *The purpose of the relatively long explanation of such practices is to indicate that scientifically invalid tests and testing practices are ingrained in the regulatory (FDA) setup, which, by any standard, should be considered criminal ignorance and incompetence in science.*

This practice of using invalid tests and testers extended to the biological side, leading to the coronavirus pandemic's disaster later explained in detail.

4. The strange aspect is that all alternate medicinal products/approaches are considered inferior as they do not follow these modern allopathic "scientific" standards and testing. They

should be no worse than modern medicinal products because current testing does not add value but false assurance. The irony is that alternative medicine approaches accept this weakness, in ignorance, rather than highlighting modern medication assessment procedures and practices as fraud. They also do not consider the chemistry/science aspect of the medications describing their products.

5. When considering medicines and medicinal products as chemicals, their quality can be assured by a slight modification of the tester, more accurately slight change in the current mindset.[83]

An interesting aspect is that most of the regulatory standards and requirements concerning testing, as described above, relate to assessing product quality and monitoring the release of actual drugs (active ingredients) from the product, such as tablets and capsules. The products provide only the convenience of drug administration and delivery into the body. Considering the problems described above, the products are not providing these benefits.

Conversely, if a patient buys the pure drug (in powder and liquid form) from a chemical store and takes the prescribed amount, s/he would obtain the same therapeutic effect as from a product perhaps better. Pure medicines (chemical compounds) obtained from chemical suppliers can be of higher quality standards than pharmaceuticals. They have well-defined quality tests and standards. Unfortunately, this approach appears to have been ignored, arguably lacking a clear understanding of the science of chemistry for medicines.

Industry must follow the FDA, or other national, guidelines and requirements as stated above to receive approval from the FDA. As explained above, the FDA requirements are hardly science-based and do not provide any assurance of the quality of products, hence safety and efficacy. The question is, why is it so?

---

[83] https://www.drug-dissolution-testing.com/blog/files/JAP2015.pdf

As per regulatory authorities, pharmaceuticals/medicine falls under the healthcare profession, in particular medical. Therefore, it would be worth critically evaluating if this profession can assess appropriately.

Chapter 16—Part 1: The Illusion of a Pandemic can be Created by the Medical Industry

Semantics and Framing in the Declaration of a Public Health Emergency

THE MAJOR FLAW in the declaration of a Public Health Emergency in 2020 was the use of the RT-PCR test to determine that a healthy person, without disease symptoms, could be described as a 'case of disease' and hence describe them as a risk to society. This was a pandemic based *only* on increased 'cases' of disease, which were simply a positive test, not on enormous numbers of deaths and hospitalisations. Hence, if you stopped testing healthy people with the RT-PCR test there was no pandemic or public health emergency.

The RT-PCR test only identifies a segment of a virus hence claiming that a segment of a virus in a healthy person represents a risk to society is a misinterpretation of the function of the RT-PCR test and this test was used in this way for the first time in 2020.

This misuse of the test and misinterpretation of the results permitted scientists to claim that healthy people were a case of disease, and they could be quarantined by the government. Yet the historical control of infectious diseases provides evidence that the presence of a virus in humans (even a whole virus) *does not indicate that you will ever get sick from this virus.*

Humans carry thousands of microorganisms around all the time, and they only become pathogenic if poor environment or host conditions exist. [1] In other words, the virus on its own is not an independent cause of disease, and in healthy environments we live in harmony with these microorganisms, and they do not have to be eradicated.

This is also the reason why no-one is responsible for the health of other people in the community, and it is a fallacy to promote a medication to the community as being your responsibility for the 'good of the community'. Vaccines were only ever a secondary measure after the risk (hospitalisations and deaths) from infectious diseases was controlled through government health policies that improved the environment and lifestyle of Australians.

Hence, vaccines have always been voluntary, and no government has provided scientific evidence to support mandating any vaccine for the control of any infectious disease in genetically diverse populations.

When adopting a strategy to prevent infectious diseases it is important to choose the preventative measure that best addresses the causal mechanisms for the disease. There is a wealth of data showing that environmental factors and host characteristics are the primary determinants of health outcomes after exposure to all infectious agents. [1][2][3]

The use of emergency powers under the declaration of a global pandemic has been based on a flawed change to the definition of a pandemic that occurred in 2009 (see Section 1.9 below) and the misuse of RT-PCR testing in healthy people to create false cases of disease that the media has used to frighten the public about a novel virus. The illusion of a 'pandemic' was then created by using mathematical modelling to produce a hypothetical estimate of deaths and illnesses due to the virus that was not based on real data.[4] Media images were then used to further frighten the public.

The flaws in the non-transparent assumptions in the mathematical modelling of this novel virus invalidate the declared public health emergency in Australia. An illusion of a pandemic has been achieved through changes to diagnostic criteria of influenza/coronavirus and extra surveillance of this novel virus in healthy people. [5—see Section 2.1] The current so-called pandemic situation is being maintained by the suppression of science through financial conflicts of interest in government, and by the media/Big Tech companies on public forums. Politicians, through the donations and lobbying system of government are promoting the information from corporate-public partnerships in the GAVI alliance and the World Economic Forum. These partnerships are influential in the design of global public health policies through the World Health Organisation. [6—Chapter3]

The Change in Definition of a 'Global Pandemic' 2009

In May 2009, the World Health Organisation (WHO) changed the definition of pandemic based on the advice provided by a small select committee that was not required to reveal their conflicts of interest until 12 months later. [31][32] This change in definition was critical to the ability for the WHO to declare a swine-flu 'pandemic' in June 2009 and then a coronavirus pandemic in March 2020.

The WHO could never have declared a public health emergency under the International Health Regulations in March 2020, without this change to the definition of a pandemic. It is this arbitrary change to the definition that has resulted in the removal of human rights globally under emergency powers that are not validated as being for a legitimate public health purpose.

The change to the definition that occurred in 2009 was the removal of the need for there to be an '*enormous number of deaths and illnesses*' to a new virus before a pandemic can be declared. This phrase was replaced with cases in the new definition and a pandemic could be declared simply if there was an 'increase in the number of cases of a disease', regardless of whether these so-called cases were serious or non-serious, or even if the 'cases' had no disease symptoms at all.

This change to the definition is critical because historically, epidemiologists and immunologists stated that cases of an infectious disease do not indicate the risk of the disease to the community after 1950 in Australia. [9]

Under the International Health Regulations, an anonymous committee was given the power to make decisions about vaccination policies in global pandemics without consultation with the Strategic Advisory Group of Experts (SAGE), the principal advisory group within the WHO for the development of policies related to vaccines and immunization strategy. [33]

The establishment of the Emergency Committee did not require any declaration of Conflicts of Interests (COI) or stakeholder representation in decisions to declare a global pandemic.

The important change to the definition that occurred in May 2009 was the removal of the need to show how severe the impact of the virus would be on the population with real data.

The Arbitrary Changes made to the Definition of a Pandemic in 2009

Before 4th May 2009:

> *An influenza pandemic may occur when a new influenza virus appears against which the human population has no immunity, resulting in epidemics worldwide with enormous numbers of deaths and illness. With the increase in global transport, as well as urbanization and overcrowded conditions, epidemics due to the new influenza virus are likely to quickly take hold around the world.* [31]

After 4th May 2009:

> *A disease epidemic occurs when there are more cases of that disease than normal. A pandemic is a*

> *worldwide epidemic of a disease. An influenza pandemic may occur when a new influenza virus appears against which the human population has no immunity .... Pandemics can be either mild or severe in the illness and death they cause, and the severity of a pandemic can change over the course of the pandemic.* [31]

Pandemic planning requires that all stakeholders agree on a common definition of what an influenza pandemic represents. The Parliamentary Assembly of the European Council (PA) believes that the changes made to the pandemic definition were highly inappropriate at a time when a major influenza infection was occurring. [31]

These changes affected disease descriptions and indicators and they were made in a non-transparent manner. It also meant that because of the Pandemic Preparedness Plans (PPPs) that locked governments into prescribed actions when a pandemic was called, authorities were constrained in their actions—even when the evidence did not match the actions they were required to implement. [34]

Once the pandemic was declared, governments had no choice but to buy up the required vaccines according to quantities and prices set in the PPP's.

Risk of Infectious Diseases in Australia not based on Cases after 1950

The risk from infectious diseases was removed in most developed countries by 1950. After this time cases of disease were no longer required to be reported because it was known that most cases did not lead to hospitalisations or deaths, that is, the serious cases of these diseases, and immunity was induced with these mild or asymptomatic cases (natural infections). Here is a summary from prominent public health officials about the risk due to infectious diseases after 1950 in Australia:

The Commonwealth of Australia Director General of Health (1913-1945) stated the decline of infectious diseases in Australia occurred at the same time as the period of sanitary reform and prior to the introduction of most vaccines. [7]

1. Professor Fiona Stanley, Australian of the Year 2003 "Infectious deaths fell before widespread vaccination was implemented." [8]
2. Australian immunologist and Nobel laureate, MacFarlane Burnet, stated in 1952 that the risk of infections to the community can only be determined by examining the age-incidence of death and illness, (hospitalisations) not the overall incidence (cases) of the disease in the population. [9]
3. The prominent UK public health official, Stewart, confirmed in the 1970's that notifications (cases) are an incomplete indicator of prevalence, and they are not an indication of the severity of the disease in the population. [10]

In Australia, measles, whooping cough (pertussis) and influenza were removed from the National Notifiable Disease list by 1950. [11] That is, doctors were no longer required to notify the government of cases of these diseases because most people did not die or get hospitalised from these diseases. Mortality and morbidity had declined by 1950 and the diseases were no longer considered diseases of serious concern to the community, even though there were still cases of these diseases. There were no vaccines for these diseases in 1950, so vaccine-created herd immunity did not remove the *risk* of these diseases.

The evidence above shows that vaccine-created herd immunity was not the major reason for the decline in risk from infectious diseases. In 2010, Professor Terry Nolan (chair of ATAGI 2005-2015) confirmed this fact by stating that demonstrating herd immunity from a vaccine was "neither necessary nor sufficient for a positive recommendation on the national immunisation program (NIP)." [12] Hence, there is no justification for mandating any vaccine in the community. Further, acceptance rates for vaccines in voluntary mass vaccination campaigns did not reach 50% even by 1980, [13] they were all voluntary, and infectious diseases have been a low risk for Australians (and in all developed countries) since 1950.

Whilst the Polio vaccine was credited with controlling the Polio disease in 1956, the medical literature shows that this situation was achieved through changes to the diagnostic and surveillance criteria of 'cases' of Polio in the population. These changes are noted to be similar in nature to the creation of an illusion of a 'pandemic' of COVID-19 disease in 2020. [14—See Part 2.3]

In addition, individuals are not equally susceptible to infectious agents due to genetics and many other factors [1][9][15] and there is a range of outcomes that can occur after infection with any virus/bacteria, including from *Coronavirus 2019*. These outcomes include no symptoms at all (subclinical infections that are asymptomatic), mild disease, severe disease, or death. Focusing on the overall incidence (cases) of infectious diseases, by publicising every case, does not inform the public of the risk of the disease in the community. That is, the deaths and serious illnesses (hospitalisations) occurring due to these infections.

Plus, asymptomatic infections create immunity in people without getting or spreading the disease. Exposure to the virus (infection) is not the same as having the disease (case). Hence, we have never considered people without symptoms to be a 'case' of disease or a risk to society *prior* to 2020.

In all developed countries public health reforms, nutrition and smaller family sizes resulted in mostly non-serious cases of infectious diseases (99.9%) after 1950, even when infection rates were high. [9][16][8][17] Death and serious disease from infectious diseases were extremely rare after this time in Australia and in all developed countries. Influenza-like illness has not been a significant risk in Australia since 1950 and this cannot be due to vaccination because a vaccine was not introduced into voluntary vaccination programs in Australia until after 1997 and its efficacy is debatable.

It is also observed that infectious diseases are common in highly vaccinated populations. [3][18] Most cases of infectious diseases in developed countries are non-serious cases of disease. They are self-limiting and the individual will receive long-term immunity from this natural infection. This is how herd immunity was originally established once enough of the population had been exposed to the infectious agent. [1][19] Hence, publicising each case of this disease in the media as if every case is a public health emergency is not informing the public of the lack of severity of most of these cases in developed countries or of the fact that these cases are *necessary* for herd immunity.

Burnet stated in 1952 that the risk of infections to the community can only be determined by examining the age-incidence of death and illness, not the overall incidence of the disease in the population. This is because childhood infectious diseases are mainly severe in children less than one year of age [9] and influenza-like illness is only serious to the elderly, over 65 years of age, with underlying co-morbidity.

This information is not made transparent in the statistics that are used by the government health departments or by the media to promote vaccines to the public. Currently the media reports 'cases' of COVID-19 disease that are simply a *positive test* without any disease symptoms, and these 'cases' (which may or may not be infections) are being used to encourage the *assumption* that it will result in high mortality and morbidity in the community—in every age group. This assumption is incorrect.

Most cases of COVID-19 disease reported by the media have *no symptoms* at all, so they are not cases of disease or even proven to be infections. They have resulted from healthy people receiving a positive RT-PCR test: a test that cannot diagnose COVID-19 disease or infection. Plus, the test has been run in Australia at a high cycle threshold (35-45 CT). At this high CT the results for cases are potentially 100% false positive and there is no standardisation of the CT in laboratories across Australia. In 2021, the CDC was recommending that the PCR test was run at 28 CT for the vaccinated and 35-45 CT for the unvaccinated [20]—what a fraud!

Most of the cases reported in the media are not cases at all or are not serious cases (not hospitalised and/or no symptoms) and would otherwise go unnoticed if they were not reported in the media—which was the case prior to 2020. Sub-clinical infections (without disease symptoms) provide longer-term immunity than artificially induced immunity from vaccination and they are not a risk to others. [9][16]

Australian National Public Health Policies must be designed on Australian National Data

Public health policies are never designed using the statistics from other countries. [6—Chapter 2]

This is because many environmental and host factors play a role in the pathogenicity and virulence of the microorganism in different countries. Hence it is essential to see how the agent behaves in each country before a declaration of a pandemic can be made. Closing the borders of a country based on mathematical modelling of an infectious agent *before* the agent has been observed to be a risk in that country, is false science. This is why the pandemic preparedness model used to predict this 'pandemic' was dumped by the US Surgeon-General in April 2020 for 'not using real data.' [4]

Governments have falsely claimed that everyone who gets exposed to this novel Coronavirus 2019 will get serious disease and this is the justification for screening healthy people for the virus. There is no scientific evidence for this claim. The only statistics that can inform governments of the *risk* of an infectious agent to the community are the hospitalisations and deaths, or case-fatality statistics in each demographic.

Medical Diagnosis of Disease

Case-fatality rates will vary greatly in different investigations because of the different criteria that can be used in diagnosing and reporting diseases and death. [5]

This is the hidden information in disease statistics. An appearance of an increase in a disease can be manufactured by a change in the surveillance of a disease and by changing the diagnostic criteria and/or reporting of the disease.

This is the case in 2020 when healthy people were screened for a new disease with an inappropriate test and when the reporting of influenza on death certificates was changed to allow COVID-19 disease to be the primary cause in patients with co-morbidity.

Diagnosis of disease is a grey area of science because criteria can be changed over time, and this can give the appearance of an increase in one disease and a decline in another. In addition, people usually die from multiple factors, therefore cause of death can be subjective on the death certificates. In 2020, Australia experienced an increase in COVID-19 disease (flu-like illness) and a decrease in influenza and pneumonia [35] and cases of COVID-19 did not require laboratory confirmation. Proof of causality by the presence of SARS-Cov-2 was not required for positive diagnosis and the presence of other viruses/bacteria was ignored.

This was also the case for the alleged global public health emergency in 2020. The WHO declared a pandemic of *Coronavirus 2019* based solely on the alleged identification of the virus in healthy people using the RT-PCR test designed by the CDC itself.

This extra surveillance of the healthy population resulted in thousands of people *without* symptoms and/or non-serious cases of disease, being used to frighten the public about a new flu-like illness that was called COVID-19 disease.

The Definition of a Case of COVID-19 Disease

The 'flu-like symptoms and neurological damage' experienced with COVID-19 disease are caused by hundreds of respiratory viruses, bacteria, and even medications and vaccines. So, unless there is a systematic investigation of the cause of these symptoms through laboratory testing and identification of the infective agents present, and knowledge of the medications/vaccines that are also present, then proof of causality for diagnosis has not been provided.

A probable case of COVID-19 is not laboratory confirmed—nor is a suspected case of COVID-19 disease. Yet these definitions are satisfactory for cases to be documented and published by the health department and media. [26] This diagnosis is being made for cases without any medical context of the vaccines/medications or laboratory identification of hundreds of viruses/bacteria that also cause the same flu-like symptoms attributed to cases of COVID-19 disease.

The proof that no laboratory confirmation of the virus is required to diagnose a case of COVID-19 disease is provided in the recently updated standardized case-definitions of COVID-19 disease in 2021. [26]

For example:

> *You don't need to have a positive test to be counted as a case of COVID-19. Anyone with certain symptoms who has spent at least 15 minutes within 6 feet of "a probable case of COVID-19," OR is a "member of an exposed risk cohort as defined by public health authorities during an outbreak or during high community transmission," and who does not have "a more likely diagnosis" is counted as a COVID-19 case.*
>
> *Any death certificate that lists COVID-19 "as*

an underlying cause of death or a significant condition contributing to death," with or without any laboratory evidence of COVID-19, is counted as a COVID-19 death.

The symptoms to be counted as a case of 'COVID-19 disease' include the acute onset or worsening of at least two of the listed symptoms or signs in this updated document. However, what is not mentioned in the document is that all the listed flu-like symptoms are caused by hundreds of other infectious and non-infectious agents, and no proof is required by the doctor to diagnose the symptoms as being caused by the SARSCOV-2 virus.

The RT- PCR test is not a Screening Tool for Diagnosing COVID-19 Disease

In 2020, the Australian government funded the health departments to obtain cases of COVID-19 in the healthy population. This was done by using the RT-PCR screening test that does not diagnose disease or infection. [25] The PCR test can only identify a segment of a virus and there are many viruses, bacteria and medications that cause the same flu-like symptoms and neurological damage as COVID-19 disease. Hence, the test does not diagnose disease because a segment of virus is not proof of the causal agent. This test is a supportive diagnostic tool that should only be used by doctors after specific symptoms appear.

This misuse of the test and misinterpretation of the test results enabled governments to claim that any positive result was 'an asymptomatic *case* of COVID-19 disease' even though identifying an infectious agent does not indicate that you will ever get sick with the disease and using the test at 35-45 CT produces potentially 100% false positives.

In addition, the Health Departments was permitting a

diagnosis of COVID-19 Disease without the virus being identified. Diagnosis could be based on suspected COVID-19 without testing for the virus. [26] For the first time in history, healthy people without disease symptoms became a case of disease in 2020, and their liberty could be removed based on a test that does not diagnose COVID-19 disease or infection without specific symptoms also being present.

Lack of Evidence that Asymptomatic People are a Risk

People without any disease symptoms are not and never have been classified as a case of disease or a risk to the community prior to 2020. This is because most people will never get disease symptoms in a developed country *even if they carry the infectious agent*. The World Health Organisation (WHO) admitted in June 2020 that they did not have any evidence to support the claim that an asymptomatic person is a risk to society just because they carry the virus. [21][22]

Therefore, locking down the healthy population is an unnecessary measure that causes more sickness than it prevents to the healthy population. [23] The WHO stated that healthy people rarely transmit the virus, and the evidence has shown that most people, 99.9% of those under 65 years of age do not die or get hospitalised from exposure to this novel Coronavirus 2019. [24]

On the contrary, they get immunity from this exposure.

Further, 89% of deaths to COVID-19 disease in Australia in 2020 had co-morbidity, with the median age of death being 86.9 years with over a third having existing chronic cardiac conditions. [24]

The WHO also claimed in March 2020 that most people would not have any immunity to this new mutated virus, however, this is a coronavirus, and these are common respiratory viruses in all populations causing the common cold and flu-like symptoms.

The correct assumption would be that most people would

have some immunity to a novel coronavirus. This has now been proven.

It is also known that antibody seroconversion is achieved by natural exposure to the infectious agent, with or without clinical symptoms. [1]

Cases without symptoms are referred to as asymptomatic infections (sub-clinical) and they result in long-term immunity in contrast to the short-term immunity obtained after a vaccine. [27][28][1] In March 2020, the WHO introduced lockdowns on the basis that 'asymptomatic cases of this flu-like illness (COVID-19 disease) were a risk to the community. The WHO did not provide any empirical evidence to support this statement.

Media articles that report these non-serious cases of disease without reporting the vaccination status or symptoms, leave the public to assume that the cases are all occurring in unvaccinated people and that they are all serious, that is, hospitalised/death.

This assumption is incorrect.

Many vaccinated children/adults are still getting these infectious diseases. [6][29]

This contradicts the claim that vaccine-created herd immunity is necessary/or that it can prevent COVID-19 disease.

The WHO's New Definition of Herd Immunity

The WHO changed the definition of herd immunity for COVID-19 disease in December 2020. [30]

The new definition states:

> *Herd immunity against COVID-19 should be achieved by protecting people through vaccination, not by exposing them to the pathogen that causes the disease.*

The WHO provides the following reason for this change.

> *WHO supports achieving 'herd immunity' through vaccination, not by allowing a disease to spread through any segment of the population, as this would result in unnecessary cases and deaths.*

This is an unsupported claim.

This definition *assumes* that having the virus always results in serious disease and this is *false*. This assumption about this virus opposes the scientific knowledge of how microorganisms cause disease as well as knowledge of the complexity of the human immune system in developing immunity.

The human immune system involves more than a rise in the antibody titre that is induced by a vaccine, and it is natural immunity that protects humans from serious illness and death to infectious agents and has done for thousands of years.

The Definition of a Vaccine and the CDC's Change to this Definition in 2021

The traditional definition of a 'vaccine' stated by the WHO is:

> *...a special preparation of antigenic material that can be used to stimulate the development of antibodies thus conferring active immunity against a specific disease or number of diseases.*[36]

Further, the World Health Organisation's definition states that:

> *...because the vaccine only contains killed or weakened forms of the virus/bacteria they do not cause disease or put you at risk of complications.*

This definition is misleading because it omits to inform consumers that the foreign proteins from the weakened virus/bacteria or the genetic recombinant particles, and the chemicals in the vaccine carrier, all present a risk for autoimmune diseases and cancer to recipients of vaccines. Many people are also pre-disposed to these illnesses due to their individual genetics.

The new COVID-19 'vaccines' were never tested to see if they could prevent disease transmission in the community. They were only tested to see if the product reduced the symptoms after the person got ill. This is the action of a treatment drug not a vaccine as defined by the WHO above. Pfizer falsely claimed on the product information 'this vaccine can prevent people from getting ill from COVID-19'.

The Centres for Disease Control and Prevention (CDC) also recognised that these products did not fulfil the WHO definition of a vaccine so they changed the definition of a vaccine on 1 September 2021 [37]: almost 12 months after they promoted these injections to the public as a product that would *create immunity* to a *specific disease* and *prevent the person* getting ill from COVID-19 disease.

The new US CDC definition states a vaccine "is a preparation that is used to stimulate the body's immune response against diseases." This definition no longer claims to create immunity to a specific disease or to prevent the disease. Hence, there is no demonstrated benefit in using this product over any other drug that stimulates the immune system to reduce the symptoms once you have the disease and as this drug contains new genetic technology that is untested, they also come with serious documented known and unknown risks that were not revealed to the public.

In addition, the new definition of herd immunity states that it should be achieved with a vaccine and '*not by exposing people to the pathogen that causes disease*' (Section 1.6 above). Yet now the CDC is saying that the so-called vaccine only reduces the symptoms *once you have the disease.*

The public has been educated to believe that a traditional vaccine only has rare side-effects, and this claim cannot be made for these preparations because they have not been tested in phase 3 clinical trials to determine the medium- and long-term effects in genetically diverse populations before they were mandated in government policies. Hence, these preparations do not fulfil the criteria of a 'vaccine' for either efficacy or safety.

The Definition of Vaccine Efficacy in Preventing Disease

The clinical trials for vaccines do not study the effects of vaccines on detectable infection rates in the population, instead they use the surrogate of seroconversion (antibody titre) in selected participants to claim that vaccines can prevent infectious diseases. [38] This is the case even though antibody titres are known to be an unreliable indicator of protection from disease. This does not suggest that vaccines do not have any benefit in reducing the transmission of the disease in the community, only that it is not accurate to describe infectious diseases as vaccine-preventable diseases when this criterion has not been proven by governments for all vaccines. [12]

Stanley Plotkin described as the 'father of world vaccinology', states that antibody titre is not a reliable indicator because we do not know precisely how antibodies work. [27] In other words, without the empirical clinical evidence in the population to demonstrate that vaccine-induced (artificial) antibody titre is protective against infection, we cannot claim that vaccines are effective in preventing them. This is because the immune system is much more complex than just an antibody titre response.

The Empirical Risks of Vaccines that are Ignored by Governments

The current belief stated by the Global Alliance for Vaccines and Immunisation (GAVI), an alliance of public-private partnerships (including the Federation of Pharmaceutical Companies) that advises the WHO on global vaccination programs, is that much of the burden of infectious diseases can be alleviated if every child, in every geographical location, has access to multiple vaccines. [3] However, this claim does not consider the influence of synergistic toxicity of vaccines, genetics, lifestyle, and environment on the health of populations.

Since 1990, there has been a 5-fold increase in chronic illness in children/adults in highly vaccinated populations globally and an exponential increase in autism that correlates directly with the expansion of government vaccination programs. [39][40][41][42]

This chronic illness includes childhood cancer, autism, autoimmune diseases, hypersensitivity (allergies), anaphylaxis, seizures, and behavioral and learning difficulties. Is this the genetic deterioration of the population that Macfarlane Burnet predicted in 1952 with the increased use of vaccines?

Vaccines contain foreign DNA from attenuated/inactivated or genetically engineered pathogens (virus-like particles) plus foreign animal and/or human DNA derived from the manufacturing process. There are two well established pathologies that can potentially develop from injecting children with DNA contaminants such as human foetal cells in the MMR vaccine or animal DNA, such as calf, chicken, or monkey, in other vaccines. [43]

These mechanisms include insertional mutagenesis in which the human foetal DNA inserts into the child DNA causing mutations that can lead to cancer and other diseases, and autoimmune diseases that are triggered by the human foetal DNA used in the manufacturing process of vaccines. Autoimmune diseases cause the child's immune system to attack his or her own body. This leads to diseases such as childhood rheumatoid arthritis, diabetes, hypersensitivity, allergies, anaphylaxis, autism, Crohn's disease, liver disease, etc.—all of which are escalating in children in countries with high vaccination rates.

These are diseases that are also listed by the pharmaceutical companies as being associated with vaccines for decades. [42] There is also significant research linking vaccines as a plausible cause of this chronic illness. [42][43][44][45] All these chronic illnesses have escalated in children since the expansion of the vaccination programs in 1990, and even though vaccines are demonstrated to be a plausible cause of this decline in health, governments globally have not investigated this correlation to the childhood vaccination program in properly designed causality studies.

This is despite the strength of an association, such as: i) in individual cases, ii) satisfaction of all nine of Bradford Hill's causality conditions possible given the setting and iii) additional strong evidence such as a linear dose-response relationship being consistent with cause and effect. Further, if vaccination policies truly are to protect human health, governments would be promoting vaccination programs based on the evidence that demonstrates an improvement in children's health outcomes. But they cannot do this because children's health and population health has *significantly declined* with the expansion of this program over 30 years.

A key factor that has led to this situation was the removal of liability from vaccine manufacturers for any harm caused by vaccines, in 1986. At the time, pharmaceutical companies were paying out millions of dollars in compensation for vaccine injuries and deaths every year. This removal of liability enabled governments to put vaccines on the market without proper safety testing under the guise of being 'life-saving products' when in fact they kill and injure millions of people every year.

Vaccines are not risk-free products. It was this decision by the US Congress to remove liability in 1986 that enabled governments globally to reverse the precautionary principle that is designed to protect human health in government policies. By reversing this principle, governments have placed the onus of proof of harmlessness for this medical intervention on the public, and not the manufacturers of the vaccines or the government. [46] This allows governments to ignore the evidence that parents and research institutions provide regarding the causal links of Adverse Events (AE's) to vaccines. The suppression of this scientific evidence is enhanced by the fact that governments have never looked for a causal link with proper surveillance of the adverse events. Industry-funded government regulators do not actively monitor the long-term adverse events to vaccines for 1-20 years after the injections are given.

In other words, the hard empirical evidence for the government claims of safety and efficacy of vaccines has never been collected. In the case of COVID-19 injections they were clearly experimental, and they should never have been promoted to the public as safe and effective without the necessary 10 years of long-term data that would define them as a vaccine. This terminology requires proof that the benefits of the injection far outweigh the risks. These products were falsely promoted to the public as vaccines and human health is now seriously at risk from this untested gene-technology mandated on the global population.

The deceptive practices of a medical-industry in synergy with governments and the media due to financial incentives have resulted in a medical tyranny with far reaching consequences for the human race.

References

1. Friis RH and Sellers TA. 2004. *Epidemiology for Public Health Practice* (3rd Ed). Massachusetts: Jones and Bartlett Publishers
2. World Health Organisation. 2005. Commission on Social Determinants of Health (CSDH). 2005. *Action on the Social Determinants of Health: Learning From Previous Experiences*
3. World Health Organisation. 2013. Immunisation Service Delivery (ISD). *Expanded Program on Immunisation (EPI)* (Updated 15th October 2013)
4. *US Surgeon General, Jerome Adams, Dumps the Gates/CDC/WHO Prediction Contagion Model*, https://edtvproductions.com/u-s-surgeon-general-adams-dumps-gates-cdc-who-predictive-contagion-model/
5. Wilyman J. 2009. *A New Strain of Influenza or a Change in Surveillance?* Australasian College of Environmental and Nutritional Medicine (ACNEM). Vol 28, No 4, Dec 2009.

https://www.vaccinationdecisions.net/wp-content/uploads/2014/02/ACNEM_Journal_Dec09.pdf

6. Wilyman J, 2015, A critical analysis of the Australian government's rationale for its vaccination policy, PhD thesis, University of Wollongong http://ro.uow.edu.au/theses/4541/
7. Cumpston JHL. 1989. Ed. Lewis MJ. *Health and Disease in Australia: A History by JHL Cumpston,* Canberra: Australian Government Publishing Service
8. Stanley FJ. 2001, Centenary Article: *Child Health Since Federation*, In Yearbook Australia 2001. Canberra: Australian Bureau of Statistics [ABS Catalogue No. 1301.0]. pp368-400
9. Burnet FM. 1952. *The Pattern of Disease in Childhood*, Australasian Annals of Medicine. 1: 2: pp93-107
10. Stewart G T. 1977, *Vaccination against Whooping Cough: Efficacy v Risks*, The Lancet. 29(Jan): pp234-237
11. Commonwealth of Australia (CoA). 1945-1986. Official Yearbook of the Commonwealth of Australia, No.37-72.
12. Nolan T. 2010. *The Australian Model of Immunisation Advice and Vaccine Funding*, Vaccine. 28 (April) Suppl 1: A76-A83
13. Feery B. 1981, *Impact of Immunisation on Disease Patterns in Australia*, Medical Journal of Australia. August. 2: 4: pp172-176
14. Humphries S and Bystrianyk R, 2013, *Dissolving Illusions: Disease, Vaccines and Forgotten History,* Summarised in the Masters of Health Magazine, October 2020 by Wilyman J, pp9-10

https://www.vaccinationdecisions.net/wp-content/uploads/2020/11/dr-Judy-Wilyman-feature-MOH-1.pdf

15. Gilbert SG, 2004, *A Small Dose of Toxicology: the health effects of common chemicals,* Florida: CRC Press
16. McKeown T, 1979, *The role of Medicine: Dream, Mirage*

or Nemesis? Oxford: Basil Blackwell.

17. Illich I, 1976, *Medical Nemesis: The Expropriation of Health.* London: Calder and Boyars L
18. Warfel JM, Zimmerman LI, Merkel TJ. 2013. *A cellular pertussis vaccines protect against disease but fail to prevent infection and transmission in a non-human primate model,* Division of Bacterial, Parasitic and Allergenic Products. Center for Biologics Evaluation and Research. US Food and Drug Administration (FDA). Bethesda MD, 20892
19. Colgrove J, 2006, *State of Immunity: The Politics of Vaccination in Twentieth Century America,* Berkeley: University of California Press
20. Public Health Laboratory Network, *Public health laboratory network guidance on nucleic acid test result interpretation for Sars-Cov-2,* version 1.2, 13 July 2020[84]
21. World Health Organisation, 2020, Dr. Maria Van Kerkhove, 8 June, (at 34.07-34.52 mins).[85]
22. WHO CheckYourFact December 2020[86]
23. Stock D. 2020, Why the COVID-19 Directives are Ineffective[87]
24. Australian Bureau of Statistics, *Covid-19 Mortality*, 28 October 2021[88] COVID-19 Mortality | Australian Bureau of Statistics (abs.gov.au)
25. Mullis K, PCR Test Inventor, The PCR test identifies segments of virus but cannot diagnose disease or whether

84 https://www.health.gov.au/sites/default/files/documents/2020/07/phln-guidance-on-nucleic-acid-test-result-interpretation-for-sars-cov-2.pdf

85 https://www.youtube.com/watch?v=Nm1kHCrcplw

86 https://www.vaccinationdecisions.net/wp-content/uploads/2021/08/WHO-Fact-Check-re-virus-transmission-210806.png

87 https://www.bitchute.com/video/4C19PlU6B8mt/

88 https://www.abs.gov.au/articles/covid-19-mortality-1

you will ever get sick from the virus[89]

26. Council of State and Territorial Epidemiologists (CSTE), 2021, Infectious Disease Committee Interim 20-ID-02, *Update to the Standardised Surveillance Case Definition and National Notification for 2019 Novel Coronavirus Disease—(COVID-19)*[90]
27. Plotkin. S.A. 2020, *Is There a Correlate of Protection for Measles Vaccine?* The Journal of Infectious Diseases, Vol 221: Iss 10: pp1571-1572: First published 1 November 2019
28. Australian Government (AG). Department of Health and Ageing. Immunise Australia Program (IAP) http://www.immunise.health.gov.au
29. Barrett S, Teffaha S, and Brown C, 2022, Analysis of COVID-19 Effectiveness claims in Australia, Commissioned by People for Safe Vaccines, 8 February [91]
30. World Health Organisation, 2020, *Coronavirus Disease—(COVID-19): Herd Immunity, Lockdowns and Covid-19*, 31 December[92]
31. Flynn P. 2010, *The handling of the H1N1 pandemic: more transparency needed*, Parliamentary Assembly Council of Europe. Social Health and Family affairs committee. United Kingdom.
32. Cohen D and Carter, 2010, *WHO and the Pandemic Flu 'Conspiracies'*, British Medical Journal. 340(June): c3257 doi: 10.1136/bmj.c257

[89] https://www.youtube.com/watch?v=-nXjE_W8TXo

[90] https://cdn.ymaws.com/www.cste.org/resource/resmgr/ps/ps2021/21-ID-01_COVID-19.pdf

[91] https://www.peopleforsafevaccines.org/post/analysis-of-covid-19-vaccine-effectiveness-australia

[92] https://www.who.int/news-room/questions-and-answers/item/herd-immunity-lockdowns-and-covid-19

33. World Health Organisation (WHO). 2009(i). WHO use of advisory bodies in responding to the influenza pandemic, Pandemic (H1N1) 2009 briefing note 19.
34. O'Dowd A, 2010, *Council of Europe Condemns 'Unjustified Scare' over Swine Flu*, British Medical Journal. 340 (June): c3033: doi:10.1136/bmj.c3033
35. Australian Bureau of Statistics, *Monthly Deaths due to Influenza and Pneumonia*
36. World Health Organisation, *Vaccines and Immunisation: What is Vaccination?*[93]
37. The Centers for Disease Control and Prevention (CDC), Vaccines and Imunisation: Definition of Terms[94]
38. Medical Products Agency (MPA). 2007. Public Assessment Report. Scientific Discussion. Afluria, suspension for injection, Influenza vaccine (split virion, inactivated). Mutual Recognition Procedure. SE/H/o485/01/E01. Sweden. June 28.
39. Australian Institute of Health and Welfare (AIHW). 2005. Child health, development and wellbeing. Australian Government:
    i) Selected Chronic Diseases Among Australia's Children. Bulletin 29. September 2005.
    ii) Chronic Diseases and Associated Risk Factors.
    iii) A Picture of Australia's Children. May. 2005 (accessed March 2006).
40. Public Health Agency of Canada. 2007. Thimerosal Updated Statement. Communicable Disease Report. 33(July): ACS-6. Canada.[95]
41. Burton D. 2003. Mercury in Medicine Report, US Congressional Record; Findings and Recommendations,

[93] https://www.who.int/news-room/questions-and-answers/item/vaccines-and-immunization-what-is-vaccination
[94] https://www.cdc.gov/vaccines/vac-gen/imz-basics.htm
[95] http://www.phac-aspc.gc.ca/publicat/ccdr-rmtc/07vol33/acs-06/

Safe Exposure Standard as Reported in Executive Summary, 20th May.[96]

42. Informed Consent Action Network (ICAN). 2017. Vaccine Safety: Introduction to Vaccine Safety Science & Policy in the United States (Version 1). ICAN website[97]
43. Deisher TA, Doan NV. Jarzyna P, 2016, *Insertional Mutagenesis and Autoimmunity Induced Disease caused by Human Fetal and Retroviral Residual Toxins in Vaccines*
44. Arumugham V, 2015, *Evidence that Food Proteins in Vaccines cause the Development of Food Allergies and its Implication for Vaccination Policies*, Journal of Developing Drugs, DOI: 10.4172/2329-6631.1000137
45. Arumugham V and Trushin MV, 2018, *Cancer immunology, bioimformatics, and chemokine evidence link vaccines contaminated with animal proteins to autoimmune disease: a detailed look at Crohn's Disease and Vitiligo*, Journal of Pharmaceutical Sciences and Research, Vol 10(8): pp 2106-2110
46. Wilyman J, 2020, *Misapplication of the Precautionary Principle has Misplaced the Burden of Proof of Vaccine Safety,* Institute of Pure and Applied Knowledge, Science, Public Health Policy and the Law, Vol 2: 23-33; November 28[98]

[96] www.aapsonline.org/vaccines/mercinmed.pdf

[97] https://www.icandecide.org/wp-content/uploads/2019/09/VaccineSafety-Version-1.0-October-2-2017-1.pdf

[98] https://cf5e727d-d02d-4d71-89ff-9fe2d3ad957f.filesusr.com/ugd/adf864_cb9f1c190ed547198bc085074466aaea.pdf

Chapter 17—Part 2: The Illusion of a Pandemic can be Created by the Medical Industry

Pandemics Created by Changes to Diagnostic and Surveillance Criteria

INFLUENZA DISEASE CANNOT be diagnosed by simply testing to identify a virus. This is because having the virus does not always lead to disease or even serious disease. In addition, there are hundreds of different bacteria, viruses and medications that cause flu-like illnesses. Tests are supportive tools used by doctors to assist with diagnosis only *after* specific symptoms appear.

Infections without symptoms (asymptomatic) have never been considered cases of disease before 2020. These asymptomatic infections produce immunity, and this is how herd immunity was established after public health infrastructure was implemented in many countries by 1950. Frank McFarlane Burnet won the Nobel Prize for his research on acquired immunity in 1960.

The COVID-19 'Pandemic' of 2020

In 2020, a new infectious disease appeared called 'COVID-19', yet it is only being diagnosed with a test. If you don't take the test the symptoms are the same as many other flu-like illnesses caused by hundreds of different viruses, bacteria, medications, and vaccines. Causality is undetermined if you look only for one virus.

A test without symptoms allows governments to claim that healthy people are a risk to society. Hence, by increasing the testing of the asymptomatic population in 2020, the government was able to create the *appearance* of a pandemic by finding cases of disease, with a test and no symptoms, and it can control this pandemic by stopping the testing of healthy people.

As of the 11 August 2022, the CDC removed the requirement to test asymptomatic people. So, governments can now choose when to end this pandemic by adopting these directives. Did you know that the Australian government (and other governments) stopped monitoring for flu and pneumonia in 2020? They only monitored deaths/cases of alleged COVID-19 disease. Therefore, it appeared flu and pneumonia had disappeared in 2020, but the Australian statistics for deaths to COVID-19 in 2020 were similar to the annual deaths to flu and pneumonia that occur every year.

Despite the knowledge that there are multiple causes of flu-like illnesses, governments were only monitoring for *one virus* in 2020, without specific symptoms being required. This was being done with a new PCR test that was developed by the US CDC *itself* and given emergency use only approval (EUO).

The Australian Health Minister, Greg Hunt, claimed in 2021 that cases of flu had plummeted due to the public health measures for the pandemic—ignoring the fact that the government stopped monitoring for flu and pneumonia viruses/bacteria in 2020.

All flu-like illnesses in 2020 became the COVID-19 disease based on a PCR test that claimed to identify only *Coronavirus 2019.* These tests were not accurately identifying this virus because their use was not standardised. They can produce many false positives when used at high amplifications.

The US CDC admitted in 2021 that its PCR test did not allow for the identification of influenza and other viruses known to cause a flu-like illness. Hence, they replaced this test in December 2021 with the Rapid Antigen Test and all positive results from this new test are called COVID-19 disease—but what is it identifying?

The definition of a case of an infectious disease prior to 2020 was:

> *...a set of standard criteria for classifying whether a person has a particular disease, syndrome, or other health condition.*

These tests are being misused as screening tools instead of supportive diagnostic tools. This medical fraud allowed governments to create the illusion of a pandemic based on 'cases of healthy people' instead of enormous numbers of the deaths—the criteria for a pandemic required prior to May 2009.

In June 2020, the World Health Organisation admitted that it did not have any evidence to claim that people without symptoms were a serious risk for COVID-19 transmission. The only evidence of a pandemic in Australia in 2020 was the huge number of cases being identified in asymptomatic (healthy) people.

This was a pandemic of testing—not disease. However, since the roll out of the so-called vaccine in 2021, all countries have a significant increase in cases, deaths, and hospitalisations allegedly from the COVID-19 disease. This COVID-19 diagnosis is being given to any patient who gets a positive test on admission to hospital—regardless of the symptoms—and governments have made the test mandatory for hospital admission.

This mandatory test is covering up all the adverse events from the 'vaccine' that resulted in increased hospitalisations and deaths in 2021-22. In other words, adverse events from the mRNA injections—the strokes, heart disease and blood clots, etc.—are being mislabelled. This is possible because the only thing required to diagnose COVID-19 disease is a positive test—not specific disease symptoms.

This lack of transparency in the diagnosis of this disease is resulting in the deception of global populations that is leading to increased death and illness worldwide. This is iatrogenic (medical) harm from a drug, that is falsely labelled a vaccine, and the harm is being attributed to a new virus.

The Therapeutic Goods Administration (TGA) states a drug/biologic is not a vaccine until it has had 10 years of data to establish that the benefits outweigh the risks. So why are doctors and governments promoting this mRNA injection as a vaccine without this data?

All humans have billions of microorganisms in us all the time, so identifying them when we don't have symptoms is not an accurate definition of a disease. It is also not accurate in predicting the risk from an infectious agent. An infection has many different outcomes in individuals, including no disease at all, and these asymptomatic infections induce natural acquired immunity that assists in building herd immunity in the community.

The claim that humans had a lack of natural immunity to this new coronavirus in 2020 by the WHO and US CDC was false and this has now been proven as the unvaccinated did not get this flu-like illness any more seriously than the vaccinated.

In fact, the vaccinated have been admitted to hospitals in much higher numbers than the unvaccinated. It is a pandemic of the vaccinated. If we stop testing for Coronavirus 2019 in asymptomatic people, then the pandemic stops. It is false science to have tested for one virus out of hundreds that can produce flu-like symptoms. The changes to surveillance and diagnosis of flu-like illness in 2020 with financial incentives have led doctors to violate their ethical guidelines that state, 'they must not use their medical knowledge to remove human rights.'

The Swine-flu 'Pandemic' in 2009

In 2009, the World Health Organisation (WHO) used the same blueprint again to create the illusion of a swine-flu pandemic. Here is a description of how this was done from an Australian perspective.

The Australian Government prioritized a new flu vaccine in 2009 for use against a novel swine-flu virus. This preventative action was notable at the time because there was little evidence in the community to suggest this new strain was any more virulent than other new strains of flu that occur regularly.

In fact, the World Health Organization (2009) stated the majority of people who contract this disease experience the milder form of influenza and recover without requiring treatment. [1] An examination of evidence provided by the Western Australian Health

Department regarding deaths to swine influenza Type A H1N1 prompted the question 'is it possible that a change in the surveillance of influenza in 2009 has resulted in the creation of hysteria over a new strain of influenza?'

Influenza is a disease that is caused by many strains of virus. These viruses spread easily, and new strains develop regularly. [2] A vaccine against influenza will only protect against one to three strains depending on the type of vaccine used. [3] For example, the current seasonal influenza vaccine protects against Type A (H1N1), Type A (H3N2) and Type B. [3] Influenza Type A (H1N1) is a strain that has been covered in influenza vaccines for many years.

The new strain of swine flu is stated to be a recombination of genetic material from human Type A H1N1, a strain of bird flu and 2 strains of pig flu. [1] The WHO states 'there are no known instances of humans getting this strain of influenza from pigs and other animals. It is also stated that this strain is not known to be endemic in pigs. [1] Yet this flu has been promoted to the public as swine flu even though it is a strain that has never been found in pigs. The public has been misinformed about this strain of influenza. The term 'swine flu' creates anxiety and fear of a disease that has come from pigs, when the official medical term for this new strain is 'Influenza Type A, H1N1, human strain.' [1]

The World Health Organisation stated in 2009 that swine-influenza Type A (H1N1) is a new virus and one to which most people have no or little immunity. [1] In a study conducted by the CDC it was shown that individuals between the ages of 18-64 had antibodies present that reacted to the swine flu virus. [4] Whilst this doesn't indicate clinical protection it does suggest that some individuals may have immunity from previous exposure to H1N1. [4] There was no reason to assume that the population will have no immunity to this new strain as it may be immunologically similar to previous H1N1 viruses. [5]

H1N1 is a strain of influenza that has been covered for many years in the seasonal influenza vaccine. Therefore, you would expect that the Australian Health Department would have mortality data for seasonal H1N1 from previous years. This is not the case. The Health Department has stated this data has not been collected in previous years or for this year—even though Type A H1N1 has been one of the most virulent and prevalent strains and regularly covered in the influenza vaccine. [3]

In 2009, the Australian Health Department changed the surveillance of influenza in the community. [6] The Department of Health suggested that the reason there is good data on the mortality associated with influenza H1N1 2009 is because of enhanced surveillance systems that were put in place specifically to monitor the pandemic. [6] Prior to 2009, influenza that was notified by GP's and laboratories was not systematically followed up or linked to hospitalization/death data to determine outcomes. [6]

In addition, post-mortem victims were not routinely tested for sub-types of influenza. [6] In previous years deaths were listed as 'influenza' and were not routinely sub-typed for the strain. [6] The Australian Health Department also stated 'hospitals were less likely to routinely test admitted patients with respiratory viruses, including pneumonia, for influenza, so in previous years many cases remained undiagnosed or were assumed to be primary bacterial infections. [6]

Yet in 2009, most cases of influenza notified by labs or GPs were followed up to see if the cases were hospitalized or resulted in death. The Australian Health Department was also systematically testing hospitalizations/deaths for H1N1. As a result, the health department is claiming that 90-95% of laboratory proven influenza cases are due to 'swine' H1N1. [6]

It is acknowledged that incidence figures for a disease can be inflated by monitoring a disease in a more systematic manner. A more sensitive or systematic test will identify cases that would previously have gone unidentified. However, a greater incidence of a disease does not always indicate greater severity to the population. [7] This is the case with a disease such as influenza which has a high incidence in the community, but epidemics are known to be mild for the majority of people. [8]

How can the public be sure that the number of deaths attributed to this new strain of 'swine' H1N1was different to the number of deaths associated with seasonal H1N1 in previous years if this testing was not being done?

These changes in surveillance mean that even though influenza Type A H1N1 has been prevalent in previous years, there is no data on the number of deaths associated with this strain in previous years because it hasn't been monitored. The Health Department also admits that it is unclear to what extent 'Swine' H1N1 infection may have contributed to the deaths it is linked with in 2009 because there are usually several infections present and, in most cases, underlying medical conditions. [6]

It is well known that disease diagnosis and cause of death is an inexact science, and it is up to the medical practitioner to state the primary cause of death. [9] The Health Department has not produced statistics that show the overall death rate for influenza to be significantly worse in 2009 than in previous years. [3] The Therapeutic Goods Administration, Australia's regulator of vaccines, stated in 2009 'the experience in Australia of the disease is mild in most cases.' [10]

The evidence presented above illustrates how different surveillance methods can enhance the incidence of disease in the community. This leaves the cause of the increase in incidence open to interpretation. For this reason, the government should be required to publicise any changes to surveillance practices whenever there is an increase in incidence reporting of a disease. This will ensure that the information the public receives can be interpreted in an open and transparent fashion that will lead to less fear and panic.

In addition, the government admits that the public has been misinformed by calling this strain swine flu but they have stated 'they are unable to control how the media reports on the Influenza A (H1N1) virus to the community.' [10] Why did the government not correct this information by stating it is not a swine flu and informing the public of its medical name? This is of significant concern when it is observed that fear is being used by the media to encourage the public to accept more and more vaccines (medication) in healthy people.

It is extremely important that we have an accurate knowledge of the short and long-term harm caused by using multiple vaccines in infants and adults. Until this science is complete, we need to assess carefully how many vaccines are necessary. A change in surveillance has a significant impact on the incidence of disease in the community and the consumers cannot make a proper assessment of the need for a vaccine without this information. This decision has serious consequences for our quality of life and for life itself.

The Polio 'Epidemic' in 1954

This same blueprint of changes to diagnostic and surveillance criteria was used in 1954 to report epidemics of Polio in many countries. This enabled governments in collaboration with the media to then credit a newly developed Polio 'vaccine' as the solution to controlling this disease.

It has since been publicised that neurological damage (poliomyelitis) is a condition that has many causal factors other than a single polio virus. Other causal factors include more than one strain of polio virus, the chemical DDT, arsenic, and components of vaccines. Infectious disease outbreaks are determined by local environmental conditions and host characteristics (see Chapter 16) and they should never be discussed without this environmental context.

The use of DDT became prolific throughout the 1940's to 1960's, and beyond. Yet, the epidemic of paralysis was blamed solely on one virus—the Poliomyelitis virus—by the media that was also used to promote a polio vaccine as the solution.

Changes to the definition of an epidemic and to the diagnostic criteria and surveillance of polio occurred after 1954, when the vaccine was introduced into the population. This resulted in the appearance of a decline in the disease, as DDT was phased out and polio vaccination campaigns were implemented.

This decline was largely a result of the manipulation of statistics due to changes in the definition of polio that had previously included both paralytic and non-paralytic cases of the disease. After 1954, the diagnostic criteria for polio were changed, and the two diagnostic examinations were spaced 60 days apart instead of 24 hours apart. This meant that all the short-term paralyses cases were no longer included in the definition of polio. This appearance of a decline in polio was further enhanced by changing the definition of an epidemic from 20/100,000 population to 35/100,000 population per year.

Prior to 1954, the surveillance of polio was also enhanced by government Health Departments through increased funding for hospitals. Diagnosing polio was financially incentivised by linking its diagnosis to hospital funding, just as they did in 2020 with diagnosing COVID-19 disease. This increased surveillance for polio prior to 1954, was removed after the vaccine was introduced.

Hence, whilst the statistics indicated that the disease called polio had declined in the US from 1955 onwards, the reality was that paralysis continued to increase: 50% from 1957-1958, and 80% by 1958-1959. The decline in polio disease was enhanced again in 1958, when non-paralytic cases of polio that showed meningeal signs were re-classified as aseptic meningitis.

These changes in diagnostic criteria and surveillance have been well documented by Dr. Suzanne Humphries, MD, in her book *Dissolving Illusions: Diseases, Vaccines and History you don't Know*.

The media has always been a tool used to control populations. Plato stated, 'Those who tell the stories rule the planet'. Was 2020 a repeat of history, and is it possible that in 2020 the neurological damage that we are seeing has been caused by the flu vaccine and other antivirals/medications that have not been reported in the context of these deaths and illnesses? Proof of causality using proper scientific method is essential before we give up our freedoms under the guise of protecting the community.

References for the Swine Flu 2009 Section

1. The World Health Organization (WHO) www.who.int/csr/disease/swineflu/frequently_asked_questions/about_disease/en/index.html (visited 17.9.09)
2. Jefferson T, Rivetti D, Di Pietrantonj C, Rivetti A, Demicheli V, 2008, Vaccines for Preventing Influenza in Healthy Adults, Cochrane Database of Systematic Reviews 2007, Issue 2. Art. No: CD001269
3. Government of Western Australia, Department of Health, Communicable Diseases Control Directorate, Influenza Fact Sheet, 2009
4. Centers for Disease Control and Prevention, 2009, Morbidity and Mortality Weekly Report (MMWR) 58, p. 521-524
5. Schuchat A, 2009, as cited in CDC, MMWR 58, p.521-524
6. Government of Western Australia, Department of Health, Communicable Diseases Control Directorate
7. Burnet, M., 1952, The Pattern of Disease in Childhood, Australasian Annals of Medicine, Vol.1, No. 2: p. 93.
8. Heikkinen T, Booy R, Campins M, Finn A, Olcen P, Peltola H, Rodrigo C, Schmitt H, Schumacher F, Teo S, Weil-Olivier C, 2006, Should Healthy Children be Vaccinated against Influenza? European Journal of Pediatrics, 165: 223-228, DOI 10.1007/s00431-005-0040-9
9. McIntyre P, 2009, Australian Government, Department of Health and Ageing, National Centre for Immunisation Research and Surveillance (NCIRS).
10. Australian Government, Department of Health and Ageing, 2009, Therapeutic Goods Association (TGA)

Chapter 18—Influenza Vaccine is not Proven Effective or Safe with Empirical Evidence

Introduction

THIS IS A case study of the Australian data describing the risk of influenza to the Australian community from the 1970's to 2010. It specifically examines the reasons and ethics for introducing a flu vaccine to the childhood vaccination program from 2008 onwards. This information is based on an assessment of the data from this time because this was the beginning of the push for coercive government influenza vaccination campaigns.

This analysis focuses on the empirical evidence that is available and not just observations from epidemiological studies because it is known that the parameters of these studies can be chosen to influence the outcomes for a pre-set agenda. It should also be noted that governments use mathematical modelling to determine the cost-benefit of vaccines to be added to vaccination programs. These models contain hidden assumptions, and this is the significance of emphasizing the empirical data presented in this assessment of influenza vaccine effectiveness and safety.

This assessment quotes the evidence from the Cochrane Database in 2008—prior to the sponsorship of the Cochrane Database by the Bill and Melinda Gates Foundation in 2018.

The independence of this institution after this date is in question because the Bill and Melinda Gates Foundation is behind the push for global vaccination campaigns through the World Health Organisation's Global Health Policy agenda.

Background

Western Australia (WA) was the first Australian State to participate in a campaign offering free influenza vaccine to children in 2008. This campaign was being conducted through the Telethon Institute for Children's Health Research and funded by the vaccine companies—CSL Laboratories and Sanofi-Pasteur at a cost of $1.2 million. [1a] The stated purpose of the trial was a pre-emptive attempt to protect young children from influenza and to assess the efficacy of the vaccine in preventing influenza in the community. [1a]

In Australia, coercion is being used to encourage parents to use 12+ vaccines (~24 doses) in infants before they are twelve months old. [1b] This analysis examines the government's evidence for the claim that influenza is a serious risk to children, and it looks at the promotional campaign that was run in WA in 2008 to determine if influenza vaccine was being promoted to parents with accurate and balanced information on the risks and benefits. Or was it based on a fear campaign run by the media? This analysis addresses whether there is strong evidence for the claim that influenza vaccine should be recommended to all demographics and whether governments have provided evidence it is safe to use multiple vaccines in infants.

The most conclusive evidence for determining the health effects of combining multiple vaccines in infant bodies comes from long-term prospective studies on animals and humans. [2] This type of study provides evidence of the cumulative, synergistic, and latent effects of the chemicals in vaccines on children's health outcomes. A search of government and medical documents shows neither of these studies has been done. [3a]

The evidence being used by advisory committees to claim it is safe to use multiple vaccines in infants comes from short-term epidemiological (statistical) studies observing one or two vaccines at a time. [2b] This does not give us the complete picture of how the combination of vaccines will affect children's health.

In the absence of conclusive evidence on the safety of childhood vaccination, data on the ecological health of Australian children should be used as a safety signal for this program. Ecological health is the overall change in health outcomes that are observed in the population of Australian children over 5-10 years. These outcomes should be evaluated to assess if there is a suggestive link between the use of multiple vaccines in children and a significant increase in chronic illnesses. This evidence is provided in this analysis.

The Risk of Influenza in Australia

Mortality data for influenza indicates that the number of deaths for children under 5 years of age in Western Australia is between zero and three deaths per year. [4] This has been the number of deaths for the last 4 decades up to 2008, and it is similar to all other Australian states. [4] National deaths from influenza in children under 5 have been between zero and three annually since 1977. [4] These statistics do not justify the general vaccination of children for influenza. [5]

It is stated in the Cochrane review of influenza vaccines that the consequences of influenza in children and adults is mainly absenteeism from school and work. [6]

The hospitalization and mortality data shows that the risk of complications and deaths from influenza is greatest in people over 65 years old and that there is an increased risk of complications from influenza in children under 2. [4]

The assessment of the risk of this disease should also include the morbidity from complications of influenza and an assessment of the social circumstances surrounding hospitalized cases.

Social conditions should be assessed with cases of this disease because the incidence of infectious diseases increases with poor living standards and other social factors such as nutrition. [7] Currently this data is not reported. [1d] It should also be noted that a decision to vaccinate all children for influenza should not be based upon data from other countries as local factors such as living conditions, nutrition, available healthcare, and patterns of childcare will affect the benefits of using the vaccine. [5] Yet the Western Australian Health Department has based its promotion of childhood influenza vaccination on data from other countries. [1b]

It is observed that the attack rates for influenza are consistently high in children during annual outbreaks. [5] However, even when the attack rates are 20-30% it is known that the majority of these children make a full recovery and discomfort is the main symptom of illness. [5]

A recent survey of US pediatricians illustrated that 43% actively opposed the universal vaccination of children and 27% were unsure. [5] In addition, 50% of pediatricians were concerned about the safety of the inactivated vaccine. [5]

Complications of influenza include acute otitis media, croup, bronchitis, pneumonia, and other respiratory diseases such as asthma and most children with these conditions are not hospitalized.

It is observed that infants under 3 years and young children with underlying medical problems are at highest risk of being hospitalized. [5] The highest risk of influenza-associated hospitalization is in infants under 6 months of age, yet the inactivated influenza vaccine is only licensed for use in children 6 months and over. [5]

Today's children receive multiple vaccines and inclusion of the influenza vaccine results in some children receiving up to 14 vaccines before five years of age. [1c] The combination of multiple vaccines must be considered when weighing up the risks of diseases, as vaccines contain antibiotics, preservatives and aluminium adjuvants that are known allergens and neurotoxins. [10]

The cumulative and synergistic effects of the increased number of vaccines must be considered. It is also necessary to determine how effective the vaccine is in preventing influenza in the community.

Safety and Efficacy of Influenza Vaccine

Influenza is a disease that is caused by hundreds of strains of virus; however, the vaccine only protects against one to three strains depending on the type of vaccine used. [1e]

The government uses two definitions to describe the effectiveness of the vaccine. The term efficacy is used to describe how well the vaccine protects against the 3 strains of influenza covered by the vaccine. For example, the current vaccine in 2008 protects against Type A (H1N1), Type A (H3N2) and Type B. [1e]

The term effectiveness of the vaccine is used to describe the ability of the vaccine to protect against 'Influenza-Like Illness' (ILI), that is, the influenza cases that are not laboratory confirmed and the strain of virus is unknown. [1e]

Therefore, some ILI will be caused by strains of virus that are not present in the vaccine and these cases will not be recorded in the surveillance of influenza. So, the only real indicator of whether vaccine programs are reducing the incidence of influenza in the community is to monitor the hospitalizations and death due to all influenza-like illness (ILI) each year—not just a percentage of cases that are sub-typed for strains covered by the vaccine. At present the WA government is reporting only on some hospitalizations that are sub-typed for the strains of influenza covered in the vaccine to support its policy. [1d]

It is possible that because there are many viruses causing influenza illness in the community reducing the circulation of 2 or 3 will not reduce ILI in the community as other strains of influenza will infect. This is another reason why it is important to analyze hospitalization and mortality data to ensure this program is achieving its outcomes. An assessment of this data will confirm whether predicting the most severe strain of influenza virus a year in advance is a successful strategy.

In Australia the flu vaccine has been offered free to people 65 years and older since 1999. This program has had an uptake rate of 79%. [4b]

The strongest evidence for the effectiveness of this campaign would be an analysis of the hospitalization data and deaths in this age group since the program started ten years ago. This analysis has not been published or presented as evidence in the formulation of current influenza policies in 2008. [3a]

A recent Cochrane Review of all the studies conducted on the effectiveness of influenza vaccines in children stated that the efficacy of inactivated vaccines for children under two (against strains contained in the vaccine) was similar to placebo, that is, not effective at all. [6]

It should also be noted that the Cochrane Review states that neither type of influenza vaccine—inactivated or weakened influenza viruses (nasal sprays) were good at preventing ILI in children over 2. [6]

It was also concluded in the Cochrane Review that due to the variability in study design an analysis of safety data for influenza vaccines in children was not feasible. [6]

Despite significant coverage of the influenza vaccine in the Australian community for many years, both in the elderly and in workplaces, 2007 was described as a severe flu season with notifications being 3.4 times the 5-year mean (3c). In Western Australia it was described as being the worst flu outbreak in four years (12). This evidence is not an indication that influenza vaccine is reducing the incidence of this disease in the community.

Although notifications for this disease are highest in the 0-4 year age group, this is not a reflection of the severity of the disease in the population. This is because influenza is only considered a serious disease in the elderly and immune compromised and the majority of children and adults make a complete recovery after several days. [5] [13] [7]

It is hospitalisations and deaths that indicate the risk of influenza, not cases of flu from which most people recover and get immunity.

The Health of Australian Children

Our knowledge of the effects of vaccines has now been collected for over 100 years. It is important to look at the ecological health of the population as well as the statistics collected over this time to ensure this procedure is safe. Statistics can hide many variables. The ecological evidence is showing that the health of children has not improved as the number of vaccines on the childhood schedule has increased. [4c] Chronic illness in children has risen dramatically in the last two decades and this coincides with the government's push to increase vaccination rates in Australia to 95% with the addition of many more vaccines to the childhood program. This was started with the implementation of the Immunise Australia Program in 1993. [3b]

Children's health and the health of society are dependent upon scientifically proven preventative policies. It has not been proven that vaccines are not a cause of chronic illness in children. The increase in autoimmune diseases in dogs and cats has already been linked to vaccines and we must consider this same possibility in our children. [14] [15]

The Evaluation and Promotion of Influenza Vaccination in Western Australia

Australia has adopted the guidelines set by the Centre for Disease Control's Advisory Committee on Immunization Practices (ACIP, USA) which state that annual vaccination of all children aged 6 months to 4 years should continue to be a primary focus of vaccination efforts because these children are at higher risk for influenza complications compared to older children. [16]

The ACIP also recommends that annual vaccination be administered to all children aged 5-18 years. [16]

Western Australia adopted this initiative in autumn 2008. The WA Health Department promoted free childhood influenza vaccines through an advertising campaign in the media. The advertisements used the deaths of three children in 2007 to suggest influenza is a serious risk to all children. [12]

Further examination of these deaths revealed that the cause of death for these three children was inconclusive and still subject to a coroner's report at the time of the media promotion. [1b]

The Director of the flu campaign stated the information on these deaths was restricted to the public, yet the information was used in a state-based media campaign. [1b] These deaths represent anecdotal evidence of the risk of influenza to children and it was revealed in the media that only one of the children was confirmed with Influenza A as opposed to all three children that the vaccine advertisement had implied. [12] All three children were confirmed to have bacterial pneumonia, but this was only provided in the fine print underneath the advertisement.

Discussion

Children are considered the main transmitters of influenza in the community. [5] Heikkinen et al (2006) therefore suggest it is logical to assume that vaccinating children would lead to substantial reductions in parental work loss due to caring for sick children. [5]

They also suggest it could lead to decreased morbidity and mortality in the elderly. This is an assumption that ignores the theory of opportunistic infection. It is possible that because there are many strains of influenza viruses circulating in the community, reducing the incidence of 2 or 3 will still leave individuals susceptible to other circulating strains that will increase.

The fact that severe outbreaks of influenza are still being observed despite vaccination campaigns would appear to support this theory. In other words, matching for 2 or 3 strains of influenza virus is not affecting the incidence of influenza-associated illness (ILI) or deaths in the community because there are many other strains that cause influenza-like infection.

Heikkinen et al (2006) state that the 'average efficacy of inactivated influenza vaccine is approximately 70-80%.' [7, p.226]

That is, it will reduce 70-80% of influenza caused by the strains of virus covered by the vaccine. These authors then imply that the vaccine will reduce rates of illness, influenza-associated complications, and hospitalizations among vaccinees by the same percentage. However, they admit that the overall effectiveness of the vaccine will be reduced substantially in the community because of the other respiratory viruses/bacteria (including other strains of influenza) in circulation. Therefore, predicting the percentage reduction of influenza in the community is not possible. [5] Empirical evidence is essential in determining the reduction in illness that will result in the community because of the many variables involved.

The evidence being used by policymakers regarding the efficacy of influenza vaccine is derived from random and quasi-randomised controlled trials and observational studies. [17] Jefferson (2006) explains that many of these studies are of poor methodological quality and are known to be affected by bias and confounding factors. [18] As a result, these studies provide inconclusive evidence on vaccine effectiveness which leaves the issue open to debate.

To provide more conclusive evidence on vaccine effectiveness in the community the government must provide the hospitalization and mortality data of all influenza-like illness monitored over the period of vaccine usage. Until this data is published the effectiveness of influenza vaccine will be unknown.

Conclusion

Influenza has been promoted to the public in WA as a serious risk to children even though the influenza-associated mortality for children is described as extremely low. The children at highest risk from influenza are children under 6 months of age and the vaccine is not licensed for this age group.

In addition, the inactivated flu vaccine has been described as ineffective for children under two—the age group with the highest complications to flu. Childhood vaccines have also been described as having low effectiveness against ILI. The other reason for vaccinating children is to see whether it lowers the transmission of influenza in the community.

Statistics can hide many variables, so it is important that the public is presented with information that best represents the severity of this disease in the community—not the incidence in the community because this disease is not severe in all age-groups. At present the government only reports on a percentage of sub-types of influenza that are hospitalized. In order to show the effectiveness of the vaccines against flu in the community the government must report on all cases of ILI that are hospitalised.

Monitoring ILI will inform us whether the theory of selecting to protect against 2 or 3 strains of influenza is effective in reducing the hospitalisations and mortality of influenza in the community: the desired outcome of vaccination programs. It is possible that targeting 2 or 3 species only allows a space for one of the many other influenza viruses/bacteria to cause infections. In this case there will be no reduction in hospitalizations due to influenza-associated illness.

Other evidence that should be used in the risk analysis of influenza vaccination is the ecological evidence in the population. In the case of influenza campaigns and children's health, there are two ecological trends that are being observed:

1) Communities are still experiencing severe outbreaks of influenza despite vaccination campaigns in the elderly and in workplaces.
2) Children's health has declined as the use of vaccines has increased. If scientists cannot prove vaccination is not the cause of chronic illness, then it is unethical to continue adding vaccines to the childhood schedule.

A decision to use influenza vaccine must also consider evidence regarding the effectiveness of the vaccine. The Cochrane systematic review of vaccines does not suggest this vaccine is effective in children—particularly those under two.

The evidence suggests the Western Australian Government has run a fear campaign in the media, based on anecdotal evidence, to encourage parents to vaccinate their children. If the government is misrepresenting the risk of influenza to children and over stating the benefits of the vaccine to the community, this policy could have serious consequences for children's health and society. It also undermines the independent nature and credibility of the government advice.

Vaccines are not without risk, so it is important that value judgments about the necessity for a vaccine are made from non-biased sources. The government must therefore be seen to be openly informing parents on this issue. It is also essential that governments consider the possibility that multiple vaccines in infants are doing more harm than good particularly as this link has been described in veterinary journals and it is known that individuals can be genetically pre-disposed to chronic illnesses.

This research has implications for the use of coercive and mandatory vaccination policies. Governments need to be more selective about the vaccines they recommend on the childhood schedule and vaccination policies should remain fully voluntary unless the government can demonstrate the community is seriously at risk if a vaccine is not used.

> *The deepest sin against the human mind is to believe things without evidence.*
> —Aldous Huxley

References

1) Government of Western Australia, Department of Health,
    a) Media Release 15th February 2008, *Free Vaccines to help Fight Child Influenza.*
    b) Communicable Diseases Control Directorate, Van Buynder P, June 2008
    c) Childhood Immunisation 2009
    d) WA Communicable Diseases Bulletin, *Disease Watch,* March 2009, Vol13, No.1
    e) Communicable Diseases Control Directorate, *Influenza Fact Sheet*, 2009
    f) WA Public Health Bill 2008
2) Friis RH and Sellers TA, 2004, *Epidemiology for Public Health Practice* (3rd Ed.), Jones and Bartlett Publishers, USA
3) Australian Government, Department of Health and Ageing,
    a) National Centre for Immunisation Research and Surveillance (NCIRS)
    b) Immunise Australia Program
    c) Australian Influenza Report, Report No.13 Week ending 13 October 2007.
4)
    a) Australian Government, Australian Institute of Health and Welfare, National Mortality Database, GRIM Book Influenza, 2005.
    b) Australian Government, Australian Institute of Health and Welfare, 2005, *Adult Vaccination Survey* October 2004: summary results, AIHW cat. No. PHE 56. Canberra: AIHW & DOHA
    c) *Child Health, Development and Wellbeing: A Picture of Australia's Children* (May, 2005) www.aihw.gov.au visited 10.03.06
5) Heikkinen T, Booy R, Campins M, Finn A, Olcen P, Peltola H, Rodrigo C, Schmitt H, Schumacher F, Teo S,

Weil-Olivier C, 2006, *Should Healthy Children be Vaccinated against Influenza?*, European Journal of Pediatrics, 165: 223-228, DOI 10.1007/s00431-005-0040-9

6) Jefferson T, Rivetti A, Harnden A, Di Pietrantonj C, Demicheli V, 2008, *Vaccines for Preventing Influenza in Healthy Children*, Cochrane Database of Systematic Reviews, Issue 2, 2008, Art. No.: CD004879.
7) Burnet, M., 1952, *The Pattern of Disease in Childhood*, Australasian Annals of Medicine, Vol.1, No. 2: p. 93.
8) Eldred BE, Dean AJ, McGuire TM, Nash AL, 2006, *Vaccine Components and Constituents: Responding to Consumer Concerns*, Medical Journal of Australia, Vol. 184 Number 4, 20th February 2006.
9) Gilbert SG, 2004, *A Small Dose of Toxicology: the Health Effects of Common Chemicals*, Boca Raton Fla, CRC Press.
10) Australian Medical Association (AMA) www.ama.com.au/FAQ visited 11.04.10.
11) Jefferson T, Rivetti D, Di Pietrantonj C, Rivetti A, Demicheli V, 2008, *Vaccines for Preventing Influenza in Healthy Adults*, Cochrane Database of Systematic Reviews 2007, Issue 2. Art. No: CD001269.
12) Western Australian Government, Dept. Health, 2008, Flu vaccination advertisement, The West Australian Newspaper, 6th July 2007. www.public.health.wa.gov.au
13) Hays JN, 2000, *The Burdens of Disease: Epidemics and Human Response in Western History*, Rutgers University Press, New Jersey/London.
14) La Rosa, W.R., 2002, *The Hayward Foundation Study on Vaccines; a Possible Etiology of Autoimmune Diseases.* www.homestead.com/vonhapsburg/haywardstudyonvaccines.html visited 18.01.06
15) O'Driscoll, 2006, *Shock to the System; The Facts about*

Animal Vvaccination, Pet Food and How to Keep Your Pets Healthy, Abbeywood Publishing Ltd, 2005, Great Britain

16) US Government, Department of Health and Human Services, Centers for disease Control and Prevention, 2008, Prevention and Control of Influenza; Recommendations of the Advisory Committee on Immunisation Practices (ACIP), Morbidity and Mortality Weekly Report (MMWR) 17th July 2008.
17) Rivetti A, Jefferson T, Thomas R, Rudin M, Rivetti A, Di Pietrantonj C, Demicheli V, 2008, *Vaccines for Preventing Influenza in the Elderly*, Cochrane Database of Systematic Reviews 2006, Issue 3. Art No.: CD004876
18) Jefferson T, 2006, *Author's Response to Influenza Vaccination: Policy v Evidence*, The British Medical Journal, Letters; 333:1172 (2 December), doi: 10. 1136/bmj

Chapter 19—How the Climate & Vaccine Fraud Is Being Beaten in the Courts

DESPITE MAINSTREAM MEDIA silence, the resistance to globalist tyranny based on junk science—either against climate and COVID-19 scaremongering—is winning in the courts.

Government scientists consistently fail to bring into courts any hard evidence for their fantastic claims humans are dangerously warming the planet or spreading a deadly virus.

In response to these phony crises, crony politicians have offered so-called solutions that always involve stealing our wealth and freedoms, driving people into depression and suicide and big profits for corporations. According to the Legatum Institute the lockdowns pushed 900,000 people into poverty in UK, while the rich got 54% richer.

A paper produced by the think tank stated that the "deterioration of the labour market" caused by the response to COVID-19 had "a significant impact on poverty." [99]

Writing in The Telegraph, Baroness Stroud of Fulham, the think tank's chief executive and former government welfare adviser during David Cameron's premiership, said that the findings showed the "significant impact of economic and social restrictions on poverty levels." [100]

"Essentially, lockdowns and restrictions caused just shy of one million people to experience poverty," the Tory peer stated, as she warned that those in poverty have a lower life expectancy.

---

[99] https://www.msn.com/en-gb/money/other/covid-lockdowns-plunged-nearly-a-million-people-into-poverty-warns-think-tank/ar-AAS8Uc4

[100] https://www.telegraph.co.uk/news/2021/12/25/patchy-covid-data-could-condemn-thous

This book's co-author, John O'Sullivan, made our case on Asia Pacific TV broadcast with Mike Ryan.[101]

You may be forgiven for believing we have been losing because you don't hear of our successes in the Fake News legacy press. But the wins have been coming steadily.

Regular readers of Principia Scientific International over the past couple of years have read our reports of court cases already settled which have demonstrated lack of evidence for the claims that carbon dioxide dangerously warms the planet, and more recently, admissions from authorities that there is no gold standard isolated from the COVID-19 virus infection.

In effect, if there is no proof of a virus then there can be no test or vaccine for it. The pandemic, like CO_2-driven global warming, was concocted within the realms of secret science jiggery-pokery. While some of the known ingredients of the COVID jabs cause biological harm, it is even more concerning that the unknown and undisclosed ingredients may present an even greater threat to human health.

On this issue, we took note of the work of Dr Robert Young and his team.[102]

They conducted research to identify the specific ingredients in the Pfizer, Moderna, AstraZeneca and Johnson & Johnson COVID-19 injections. On 20 August, they published their findings. Their paper concludes that these COVID-19 injections…

> *…are NOT vaccines but nanotechnological drugs working as a genetic therapy…All these so-called "vaccines" are patented and therefore their actual content is kept secret even to the buyers, who, of course, are using taxpayers' money. So, consumers (taxpayers) have no information about what they are receiving in their bodies by*

[101] https://open.spotify.com/episode/2kefYiQmOC0I70cDeOePls

[102] https://www.drrobertyoung.com/meet-dr-young

> *inoculation.*[103]

Dr Young discussed his findings in depth during an interview which you can watch in an important Bitchute video.[104]

Would you like to see the raw data that produced the "90% and 95% effective" claims touted in the news? We would, too. But the vaccine companies will not let us see that data. As pointed out in the BMJ, something about the Pfizer and Moderna efficacy claims smells really funny.[105]

There were…

> *…3,410 total cases of suspected, but unconfirmed covid-19 in the overall study population, 1,594 occurred in the vaccine group vs. 1,816 in the placebo group.*

As Christian Elliot wrote:

> *Did they fail to do science in their scientific study by not verifying a major variable? Could they not test those "suspected but unconfirmed" cases to find out if they had covid?*
>
> *Apparently not. Why not test all 3,410 participants for the sake of accuracy? Can we only guess they didn't test because it would mess up their "90-95% effective" claims? Where's the FDA?*
>
> *Would it not be prudent for the FDA, to expect (demand) that the vaccine makers test*

---

[103] https://www.drrobertyoung.com/post/transmission-electron-microscopy-reveals-graphene-oxide-in-cov-19-vaccines

[104] https://www.bitchute.com/video/Z2sAH0Woz38r/

[105] https://blogs.bmj.com/bmj/2021/01/04/peter-doshi-pfizer-and-modernas-95-effective

> *people who have "covid-like symptoms," and release their raw data so outside, third-parties could examine how the manufacturers justified the numbers? I mean it's only every citizen of the world we're trying to get to take these experimental products...*
>
> *Why did the FDA not require that? Isn't that the entire purpose of the FDA anyway? Good question.*[106]

Christian Elliot also rightly points to the data gaps in the 'secret science.' When vaccine makers submitted their papers to the FDA for the Emergency Use Authorization,[107] among the many data gaps they reported was that they have nothing in their trials to suggest they overcame that pesky problem of Vaccine Enhanced Disease.[108]

They simply don't know, i.e., they have no idea if the vaccines they've made will also produce the same cytokine storm (and deaths) as previous attempts at such products.

As Joseph Mercola points out...

Previous attempts to develop an mRNA-based drug using lipid nanoparticles failed and had to be abandoned because when the dose was too low, the drug had no effect, and when dosed too high, the drug became too toxic. An obvious question is: What has changed that now makes this technology safe enough for mass use?[109]

---

[106] https://www.deconstructingconventional.com/post/18-reason-i-won-t-be-getting-a-covid-vaccine?postId=4b6beceb-3fa2-45d4-9e95-3b96defba00b

[107] Note: An EUA is *not* the same as a full FDA approval

[108] https://www.fda.gov/drugs/types-applications/investigational-new-drug-ind-applica

[109] https://articles.mercola.com/sites/articles/archive/2021/03/08/pfizer-covid-vacci

From hard bitten experience, Principia Scientific (PSI) colleagues backed co-founder, Dr Tim Ball, in his eight-year court battle with climate fraudster, Michael Mann, we know it takes lawyers, deep pockets and honest judges and jurors to expose these secret science scams.

After two years of making up the science as they go along, the politicians and their secret scientists are being called to account by the steadily awakening masses.

Now as we enter the end game of the fraud, we see government 'experts' called to the front lines to defend their crumbling narratives.

Still clutching to their appeals to authority, they are popping up on social media demanding urgent clamp downs on whistleblowers exposing their mendacity. A timely case in point occurred a few days ago.

The once golden boy of the climate cult, Dr Michael Mann, beaten in the Supreme Court of British Columbia two years ago by my aforesaid PSI buddy, last week demanded even heavier censorship to try to hide the failed consensus narratives over climate and COVID.

In a statement to Forbes, Dr Mann said:

> *It's great that YouTube is taking action to stop the spread of misinformation about Covid-19. But the disinformation promoted by fossil fuel-funded climate change deniers is just as rampant, and ultimately more deadly, given the profound damage we're already seeing from human-caused climate change. YouTube needs to take action here as well.*[110]

It seems that Twitter welcomed Mann's request because they

---

[110] https://www.forbes.com/sites/davidrvetter/2021/12/07/youtube-is-serving-up-climat

banned Dr Robert Malone, a key contributor to mRNA vaccine technology and an outspoken critic of COVID-19 mandates and rules. Writing on his Substack page, Malone, who had massed more than 500,000 followers, confirmed that his account was permanently suspended from Twitter and said, "*We all knew it would happen eventually.*"[111] [112]

The odious Dr Mann has built a rewarding career around secret science, which no one can see but which can always be relied on by corrupt policymakers to fit an agenda. The UK government has its own iconic figure trotting out rotten Mann-made secret science.

Most countries obliged "expert" Ferguson's lockdown mantra but he was later exposed in the House of Lords for playing fast and loose with the numbers, thanks to sterling work by Viscount Ridley.[113]

But Ferguson and Mann are like bad pennies.

As an example, as far back as November 2020, Principia Scientific reported that Health Canada has no Record of COVID-19 Virus Isolation.[114]

Freedom of Information requests to the CDC/FDA,UK and Irish health authorities also revealed governments worldwide

---

[111] https://rwmalonemd.substack.com/p/permanently-suspended-on-twitter

[112] https://www.theepochtimes.com/twitter-suspends-key-mrna-vaccine-contributor-dr-robert-malone_4183968.html?utm_source=newsnoe&utm_campaign=breaking-2021-12-29-4&utm_medium=email&est=QxrTRBDF1LKaKSijsu22WkzhcGPV0MWoCBP%2BfULGpC8WMdExsHdIcUINzHO9qO3EHG6hkyRIjVCz

[113] https://www.aier.org/article/the-failure-of-imperial-college-modeling-is-far-worse-than-we-knew/

[114] https://principia-scientific.com/health-canada-has-no-record-of-covid-19-virus-isolation/

cannot prove any such deadly coronavirus is real.[115] [116]

Same for New Zealand where Ministry of Health and NZ's Institute of Environmental Science and Research admitted to having no record of isolated COVID-19.[117]

Few have understood that our governments shield Big Pharma vaccine peddlers from liability.[118]

Vaccine peddling is the only industry in the world that bears no liability for injuries or deaths resulting from their products. First established in the U.S. in 1986 under the National Childhood Vaccine Injury Act, and reinforced by the PREP Act, vaccine makers cannot be sued, even if they are shown to be negligent.

The peddlers of the jabs have been permitted by governments to create a one-size-fits-all product, with no testing on sub-populations (i.e. people with specific health conditions), and yet they are unwilling to accept any responsibility for any adverse events or deaths their products cause.

We argue that if these corporations are not willing to stand behind their product as safe, especially one they rushed to market and skipped animal trials on, we are not willing to take a chance on their product. In short: No liability? No trust.

Secrecy is the key. Recently, Pfizer sought to have their vaccine injury data sealed by court order sealed for 55 years so no independent scientific examination which would expose their criminality to jurors in open courtrooms until we were all long

---

[115] https://principia-scientific.com/cdc-fda-confess-they-had-no-virus-they-created-the-test-for/

[116] https://principia-scientific.com/top-govt-health-official-admits-in-canada-court-covid19-virus-not-proven/

[117] https://www.fluoridefreepeel.ca/new-zealand-no-record-of-covid-19-virus-isolation-at-the-ministry-of-health-or-the-institute-of-environmental-science-and-research/

[118] https://aaronsiri.substack.com/p/covid-19-vaccine-manufacturers-can

dead, businesses destroyed, and society irrevocably altered.[119]

But Big Pharma refuses to show us the data that shows if these jabs are safe and effective. Ideally, professional journalists should be all over this mega story. They are not. Citizens are doing it for themselves.

Highly worrying is that the four major companies peddling these COVID vaccines are/have either:

1. Never brought a vaccine to market before COVID (Moderna).
2. Are serial felons (Pfizer, Astra Zeneca, and Johnson and Johnson).

For example, Moderna had been trying to Modernize our RNA (thus the company name)—for years, but had never successfully brought ANY product to market—how nice for them to get a major cash infusion from the government to keep trying.[120] [121]

Most people do not know that all major vaccine makers (save Moderna) have paid out tens of billions of dollars in damages for other products they brought to market when they knew they caused injuries and death—see Vioxx,[122] Bextra,[123] Celebrex,[124]

---

[119] https://principia-scientific.com/fda-asks-judge-to-seal-pfizers-covid-vax-safety-data-for-55-years/

[120] https://dev.modernatx.com/mrna-technology/mrna-platform-enabling-drug-discovery-development

[121] https://www.cnn.com/2020/08/11/health/moderna-vaccine-government-deal/index.html

[122] https://www.vaildaily.com/news/merck-found-liable-ordered-to-pay-damages-in-vioxx

[123] https://edition.cnn.com/2009/BUSINESS/09/02/Pfizer.fine/index.html

[124] https://medicalxpress.com/news/2021-03-fda-panel-pfizer-arthritis-drug.html

Thaalidomide,[125] and opiods[126] as a few examples.

Bear in mind the following:

1. Johnson & Johnson has lost major lawsuits in 1995, 1996, 2001, 2010, 2011, 2016, 2019.[127]

2. Pfizer has been hit with more criminal fines than any company in history. You can check out their rap sheet.[128] No wonder they are demanding that countries where they don't have liability protection put up collateral to cover vaccine-injury lawsuits.[129]

3. Astra Zeneca has also lost many lawsuits.[130]

4. Astra Zeneca had their COVID vaccine suspended in at least 18 countries over concerns of blood clots,[131] and they completely botched their meeting with the FDA with numbers from their study that didn't match.[132]

5. While Johnson and Johnson (whose vaccine *is* approved

---

[125] https://www.helix.northwestern.edu/2009/07/28/the-thalidomide-tragedy-lessons-for-drug-safety-and-regulation/
[126] https://www.nytimes.com/2019/08/26/health/oklahoma-opioids-johnson-and-johnson.html
[127] https://childrenshealthdefense.org/defender/johnson-johnson-why-trust-vaccine/
[128] https://www.mp-22.com/vax
[129] https://www.wionews.com/world/how-pfizer-tried-to-bully-argentina-and-brazil-in-exchange-for-vaccines-366037
[130] https://www.justice.gov/opa/pr/pharmaceutical-giant-astrazeneca-pay-520-million-label-drug-marketing
[131] https://www.businessinsider.com/astrazeneca-covid-vaccine-countries-suspend-denmark-thailand-batch-blood-clots-2021-3
[132] https://thehighwire.com/videos/astrazeneca-vaccine-falls-from-grace/

> for Emergency Use in the US)[133] and AstraZenca, whose vaccine is *not* approved for Emergency Use had severe issues over ingredients in over 15 million batches.[134]

You have to ask yourself, would you risk driving a car or flying on a plane where the company concerned has a special exemption if they cause your severe injury of death? Knowing such facts, why would you put these dangerous potions in your body?

Most citizens merely do as they are told and do not apply any rational critical examination on this matter. Peaceful opposition within the law requires capitalising on our enemy's weakness—their dodgy science.

We go into more of the legal implications in a later chapter.

---

[133] https://www.msn.com/en-us/news/us/fda-grants-emergency-authorization-to-johnson-johnson-vaccine/ar-BB1e4Skt

[134] https://childrenshealthdefense.org/defender/johnson-johnson-astrazeneca-covid-vaccine-ingredient-mix-up/

Chapter 20—Those Mad Mask Mandates

WHAT IS MOST saddening to those of us who work as science professionals is that throughout the pandemic, very little legitimate government science was carried out to determine what worked and what did not. Too often it has been left to independent researchers to derive clarity from chaos.

A good case in point being the insane mask mandates. We should now know definitively what the impacts of masks in our schools are. This is because authorities could have readily collected the data in large-scale, well designed and controlled experiments. Yet, despite the entire power of the state being available, nothing of the sort has appeared.

However, they did the 'masks in schools' study during autumn 2020. It showed that masks had no impact on infection levels.

Of course, our colleagues at Principia Scientific International were not slow to see what the mask madness was really all about. We have confirmation when we heard '*Masks were a softening up exercise for Plan B,*' according to a UK government whistleblower. It turns out that while there is little appetite in the British Cabinet for a full lockdown, COVID Passes are 'oven-baked' and ready to go. The masks were a visible means to readily verify how compliant the minds of Brits were.

The winter of 2021-2022 saw parliamentary division as Conservative MP's rebelled *en masse* and voted against their own Prime Minister.

Ironically, only the backing of the Opposition from Labour MP's got Boris Johnson over the line and kept on track another bleak season of heavy social restrictions. Vaccine passports voted through, 369 votes to 126. MPs also voted through legislation that requires NHS and social care staff to be vaccinated by April 2022 or face being sacked. That's a rather counterproductive way to protect the NHS.

Little do many people know, but former Prime Minister Johnson admits he is a eugenicist. In a piece written in 2007, the Prime Minister—who was formerly the Mayor of London—argued that global over population is the real issue and that the primary challenge facing the human species is the *reproduction of our species itself.*[135]

This indicates Britain, like many nations, is becoming a totalitarian police cult based on a death cult.

As Laura Dodsworth set out in her book *A State of Fear: How the UK government Weaponised Fear during the Covid-19 Pandemic*[136] masks are a nudge, even described as a *signal* by David Halpern, the director of the UK government's Behavioural Insights Team.

Similarly, Professor Neil Ferguson said that masks remind us *we're not completely out of the woods yet.*

As Dodsworth wrote:

> *'Masks were a softening up exercise for Plan B,' ...according to a government whistleblower. He told me that while there is little appetite in the Cabinet for a full lockdown, Covid Passes are 'oven-baked' and ready to go.*[137]

According to Dodsworth, the seasoned government insider plays a key role on a COVID task force and felt compelled to reveal the truth because he was so disturbed by the unethical reasons for mandating masks:

---

[135] https://freedombeacon.com/boris-johnson-once-argued-for-population-control/

[136] https://www.amazon.co.uk/State-Fear-government-weaponised-Covid-19/dp/1780667205/

[137] https://principia-scientific.com/masks-were-to-soften-you-up-for-plan-b/

> *It's a highly political move…the one-way systems, plexiglass screens and masks are to give you an illusion of the government doing something. It's just theatre. There is no evidence base or proportionality in favour of masks.*

COVID is a cult and not dependent on sound, empirical science. It proliferates because of an absence of scientific knowledge among the population—as a society we have become too lazy to verify anything for ourselves anymore. If verification requires more than a few clicks on our smartphones, we disengage and return to ignorant bliss.

Worse yet, when medical doctors see vaccine injuries inflicted on their patients and report their concerns to authorities, they are being ignored.

For example, U.S. attorney, Aaron Siri, the Managing Partner of Siri & Glimstad LLP and who has extensive experience in complex civil litigation matters, has been building a wealth of witness testimonies from medical professionals with firsthand accounts of authorities being indifferent or hostile to such whistleblowers.

Siri shared the story of the brave Dr Patricia Lee, a fully vaccinated intensive care unit physician and surgeon, who stepped forward after witnessing numerous serious injuries in her patients following COVID-19 vaccination.

Siri said:

> *Since that story was published, Dr Lee had a Zoom meeting with six federal health officials, including Dr Peter Marks of the FDA and Dr Tom Shimabukuro of the CDC. The meeting left Dr Lee more frustrated than she had been prior to the meeting—the officials had no interest in the specific harms she detailed, even after hearing the heartbreaking and traumatic stories that Dr Lee*

> *shared about her individual patients. Without asking a single question or reviewing any of the relevant medical records, Dr Lee was effectively told that COVID-19 vaccines did not cause these injuries in her patients.*[138]

Making these disturbing findings public Siri and Dr Lee—but seeing the lack of interest in them—they are upping the ante.

Dr Lee asked that both the FDA and the CDC publicly announce that physicians should not be retaliated against for advocating for their injured patients, so that they would not fear reporting injuries from COVID-19 vaccinations. Dr Lee has now reached out to the oversight committees in Congress.

Moreover, Aaron Siri, too, is pressing ahead in his legal efforts to put an end to the complacency and apparent cover up of compelling evidence of the harm vaccines are causing. Like other lawyers Siri is disturbed as to why the FDA seeks to delay for 75+ years full production of Pfizer's pre-licensure safety data.

Siri lamented:

> *While we have that fight, we submitted a request to the CDC, on behalf of ICAN, for the deidentified post-licensure safety data for the Covid-19 vaccines in the CDC's v-safe system. Even though this data is available in deidentified form (meaning, it includes no personal health information), the CDC refused to produce this data claiming it is not deidentified.*[139]

Foot dragging and obstinacy by gatekeepers in government seems to be endemic when it comes to data on vaccines injuries.

---

[138] https://aaronsiri.substack.com/p/whistleblower-fda-and-cdc-ignore
[139] https://aaronsiri.substack.com/p/fda-doubles-down-asks-federal-judge

Siri noted that:

> *A minimum of 20,010 days (54 years and 10 months). That is how long the FDA proposes to take, at a rate of 500 pages per month, to produce only a portion of the documents in its file for the COVID-19 Pfizer vaccine that PHMPT requested pursuant to the Freedom of Information Act (the "FOIA Request") and 21 C.F.R. § 601.51(e). But when it came to reviewing those same documents to license this product so that Pfizer could freely sell it to the public, the FDA took just 108 days. It took the FDA's parent department even less time to grant Pfizer complete immunity to liability for injuries from this product, and it took a stroke of the President's pen to mandate this product for federal employees, the private sector and military personnel.*[140]

A key legal point made by Siri is that with the federal government mandating that millions of people be injected with a liability-free vaccine then it requires complete government transparency—not the government's suppression of information. An important paper published by the prestigious British Medical Journal (BMJ) on this issue concluded:

> *The lack of adequate transparency about COVID-19 vaccine trials and their regulation cannot be dismissed as unfortunate, stubborn problems emblematic of the present culture in biomedicine. In a time of increasing public scrutiny, transparency of regulatory decision making leading to the approval of drug treatments and vaccines*

[140] Ibid.

> *for COVID-19 is important to ensure patient and stakeholder trust. It is a scientific, moral and ethical imperative that access to complete trial data of these global public health interventions is urgently granted to patients, researchers and other key stakeholders.*
>
> *Historically, there has been no consumer product that the federal government has mandated Americans to receive. Now, it has mandated Pfizer's vaccine to private sector employees, federal employees, the military, and more.*[141]

Relating to this, Siri added:

> *There has never been such a large-scale mandate of any product for society, let alone one that is injected into people. Even school mandates under state laws have almost always included an easy to obtain exemption. The current inability to say "no" to injecting a product into one's body absent serious consequences dictated by the government is truly unprecedented.*[142]

The sports world has mostly mandated vaccines and athletes have been forced to either abandon their careers or get the jab. At the time of writing, nearly 300 athletes have experienced cardiac arrest and over 167 have died.[143]

But not all elite athletes are toeing the line. Seeing the adverse reactions among fellow sports stars some are refusing to take the

---

[141] https://ebm.bmj.com/content/ebmed/early/2021/08/08/bmjebm-2021-111735.full.pdf

[142] Ibid.

[143] https://goodsciencing.com/covid/71-athletes-suffer-cardiac-arrest-26-die-after-covid-shot/

jab. Notably, top men's tennis player, Novak Djokovic decided 'to skip Australian Open after medical exemption denied' according to Serbian media. His stance is, in itself, a Major Championship trophy and Djokovic is now a role model for aspiring critical thinkers.[144]

The cameras at sports events pull away when the elite star crumples to the ground clutching their chest. "Unspecified medical issues" becomes the label applied whenever a sports star suffers any such adverse vaccine reaction.

In England, the home of Premier League soccer, players are beginning to speak out about their concerns. Former England International footballer, Matt Le Tissier, was the first to sound the alarm over football players suffering from heart problems. He demanded some answers as cases start going through the roof and noted that he never once saw any player suffer any heart complications during his 17-year long career.

In the recent European Championships, former Tottenham and Denmark star Christian Eriksen collapsed on the pitch after suffering a cardiac arrest. He needed an implanted heart defibrillator. Manchester United centre-back Victor Lindelof now has to wear a heart monitor. Manchester City legend Sergio Aguero was forced to retire from the sport prematurely after being diagnosed with cardiac arrhythmia following a game with Barcelona when hit by severe breathing difficulties.

[144] https://www.express.co.uk/sport/tennis/1542300/Novak-Djokovic-Australian-Open-medical-exemption-denied-Grand-Slam-tennis-news

Chapter 21—Science in the Medical Profession

A QUICK SEARCH on Google for the definition of medical professionals and physicians, provided below, clearly shows no mention of the word science or scientist associated with the professions.

Therefore, the medical subject should not be classified as a science subject and a practicing physician as a scientist. As per the definitions, the medical profession is a trade, and physicians its tradespersons trained accordingly.

A medical professional is:

- A physician or other person authorized by the applicable law to prescribe drugs in this state or another state.[145]
- A health professional (or healthcare professional) may provide health care treatment and advice based on formal training and experience.[146]
- Health professionals maintain health in humans through the application of the principles and procedures of evidence-based medicine and caring.[147]
- A medical professional is a qualified doctor who abides and is fully committed to the ethical principles and values of the medical profession.[148]

---

[145] https://www.lawinsider.com/dictionary/medical-professional

[146] https://en.wikipedia.org/wiki/Health_professional

[147] https://www.ncbi.nlm.nih.gov/books/NBK298950/

[148] https://tinyurl.com/4peuwzcm

A physician is:

- A physician, medical practitioner, medical doctor, or simply doctor is a professional who practices medicine, which is concerned with promoting, maintaining, or restoring health through the study, diagnosis, prognosis, and treatment of disease, injury, and other physical and mental impairments.[149]
- A physician is a general term for a doctor who has earned a medical degree. Physicians work to maintain, promote, and restore health by studying, diagnosing, and treating injuries and diseases.[150]

However, medical experts and physicians are often commonly assumed and customarily promoted as science experts or scientists.

It could easily be established that this indulgence of medical professionals in science, without training in science and its methodologies, has resulted in enormous problems for the healthcare system in developing and manufacturing pharmaceutical products.

Some examples of defining a scientist from the literature (Google search) are provided below.

A scientist is:

- A person who is studying or has expert knowledge of one or more of the natural or physical sciences.

A research scientist is (definitions from Oxford Languages):

- An expert in science, especially one of the physical or

---

[149] https://en.wikipedia.org/wiki/Physician

[150] https://www.webmd.com/a-to-z-guides/what-is-physician

natural sciences.[151]

- A scientist is someone who has studied science and whose job is to teach or do research in science.[152]
- A person who is engaged in and has expert knowledge of science, especially biological or physical science.[153]

How do science and its experimentation work? The scientific processes focus only on the natural world. Anything that is considered supernatural or abstract does not fit into the definition of science. Science is a systematic and logical approach to discovering and establishing natural processes. A critical aspect in this regard is that science aims for measurable results through testing and analysis.

Two critical underpinnings of the scientific method are:

(1) The hypothesis must be testable. An experiment should include a dependent variable (which does not change) and an independent variable (which does change);

(2) An investigation should include an experimental group and a control group. The control group is compared against the experimental group. Scientists, when conducting research, use scientific methods and experiments to collect measurable and empirical evidence.

The first crucial step is identifying questions and generating possible answers (hypotheses). The steps of the scientific method are often like this:

Conduct and reproduce the experiments until there are agreements between observations and theory. The reproducibility of published experiments is the foundation of science. No

---

[151] https://www.dictionary.com/browse/scientist
[152] https://www.collinsdictionary.com/dictionary/english/scientist
[153] https://www.yourdictionary.com/scientist

reproducibility—no science.

Concerning the coronavirus pandemic, the following would describe the claims (hypotheses) requiring experimental/scientific evidence:

1. There exists a virus called SARS-CoV-2.
2. The virus causes illness (respiratory infection, COVID-19) which could cause death.
3. The virus spreads from person to person.
4. The PCR test monitors the virus.
5. Covering the face with a face mask protects against virus spread.
6. Keeping a distance (approximately 6 feet) between people protects against virus spread.
7. A vaccine is needed as a treatment considering the viral nature of the infection.
8. A vaccine has been developed for protection from the virus.

However, there is no logical or legitimate scientific/experimental evidence provided in support of the claims made, e.g.:

1. Claims have been made that the virus exists and has been isolated. However, no physical sample or specimen of the virus has been provided or available anywhere in the world. Therefore, the claim is scientifically false and invalid.
2. One of the critical requirements is that study/testing must be conducted in parallel with a control group. The claims of virus existence are based on studies without parallel control groups. Study inferences are primarily based on testing a single person—no reproducibility aspect has been considered—violating the scientific principle.
3. The word isolation has been used to describe "virus isolate," not the actual "isolated virus" that linguistically

and scientifically represents two different items. This misrepresentation reflects a clear lack of understanding of scientific principles at the basic level.

4. The claim that the virus has been confirmed by establishing the RNA sequence of the virus. However, no empirical link between the two has been provided, i.e., virus and RNA sequence. Therefore, the RNA sequence-based virus identification remains an opinion, not a science-based fact.
5. The images of the SARS-CoV-2 virus (a ball with spikes) frequently shown in the literature are computer-generated images, not pictures of an actual virus. Therefore, such practice should be considered scientific deception and fraud.
6. No experimental evidence is available showing the virus's spread from person to person.
7. No experimental evidence supports the claims that face masks protect from the virus spread.
8. Promoted studies with aerosol or droplets without the virus's actual presence or use to monitor masks' efficacy in protecting from virus transfer and spread is illogical thinking and false science.
9. There has been no scientific evidence supporting the new or novel disease narrative other than opinions. The illness is commonly identified with flu-like symptoms. The science and its methods require that the illness must be clearly defined with one or more measurable and quantifiable parameters.
10. It has been suggested that illness is a respiratory tract infection caused by the virus. No measurable or physical link to the infection with the virus has been provided. It remains an opinion, not a scientific observation or fact. The suggested cause of the virus is based on opinions as no virus has been isolated from any patient.
11. No evidence or experimental testing has been provided to dismiss any other potential infection causes.

12. Routine PCR tests are conducted to establish the virus's presence or infection without appropriate validation for the virus or its RNA. Scientifically, results from any test lacking proper validation cannot be accepted. This practice clearly shows a lack of understanding of fundamental scientific principles. Therefore, all PCR test and antigen test results are scientifically false and must be dismissed—without exceptions!
13. The suggestion of a vaccine to treat the virus is scientifically invalid. How could a treatment against the virus be developed and recommended when there is no virus specimen available and no valid test for virus detection?
14. However, magically vaccines have been developed—a treatment (vaccine) without virus specimen and valid test. The development of vaccines could be considered the height of ignorance and incompetency in the subject (science) or the highest level of modern-day scientific fraud.

Medical professionals, particularly those associated with the CDC, often defend their thinking and studies as scientific. For example, the CDC highlights its mission (from the CDC website) as "To promote quality, integrity, and innovation of CDC science to improve public health." [154] Note the terminology used is "CDC science," not the science. This twisting of the words certainly does not appear to be a typo, but is intended as the title states in bold, *The Foundation and Future of CDC Science*, and appears in multiple other places, such as "The Science Support is responsible for increasing the impact of CDC science." [155]

The following part could be considered a vigorous discussion against the devious narrative of the virus and COVID-19 story.

[154] https://www.cdc.gov/os/index.htm
[155] https://www.cdc.gov/os/quality/index.htm

However, reading it standalone would be like arguing fake science with fake science. Hence one would not understand the underlying problems. Such back and forth arguments have been happening in the medicines area for at least a few decades without resolution, adding even more fake science concepts such as equating virus isolation with virus isolate.

The isolation of the viruses and their characterization (sequencing and PCR testing), among others, is an excellent example of fake science practices. Therefore, it is necessary to read the previous parts of the actual applicable and relevant science (i.e., chemistry). The medical experts do not study, understand and use it in any detail, particularly at the regulatory bodies. The authorities, in this case, CDC, developed and promoted its version of science called CDC science (the pseudoscience).

The following detailed narrative perfectly explains why CDC claims would not meet the commonly accepted scientific requirements and standards but arbitrary and self-proclaimed claims and practices of "CDC science."

Chapter 22—The Confusion of CDC Science vs. Actual Science

THE CORONAVIRUS ILLNESS or pandemic started with a piece of news or a rumor. In this case, someone in the field of medicines, particularly virology, either created or believed in the story, i.e., a new disease has developed and everyone started following it.[156]

For example:

> *WHO Traces Chinese Illness to New Coronavirus: The World Health Organization (WHO) announced a potential cause of mysterious cases of pneumonia in the Chinese city of Wuhan as stemming from a new coronavirus. WHO said the Chinese government is confirming the explanation, but the government has not yet made an official statement, reported STAT. The illness has infected at least 59 people.*
>
> *January 9, 2020—WHO Announces Mysterious Coronavirus-Related Pneumonia in Wuhan, China: At this point, the World Health Organization (WHO) still has doubts about the roots of what would become the COVID-19 pandemic, noting that the spate of pneumonia-like cases in Wuhan could have stemmed from a new coronavirus. There are 59 cases so far, and travel precautions are already at the forefront of experts' concerns.*

---

[156] https://www.ajmc.com/view/a-timeline-of-covid19-developments-in-2020

January 20, 2020—CDC Says 3 US Airports Will Begin Screening for Coronavirus: Three additional cases of what is now the 2019 novel coronavirus are reported in Thailand and Japan, causing the CDC to begin screenings at JFK International, San Francisco International, and Los Angeles International airports. These airports are picked because flights between Wuhan and the United States bring most passengers through them.

January 21, 2020—Chinese Scientist Confirms COVID-19 Human Transmission At this point, the 2019 novel coronavirus has killed 4 and infected more than 200 in China, before Zhong Nanshan, MD, finally confirms it can be transmitted from person to person. However, the WHO is still unsure of the necessity of declaring a public health emergency.

January 31, 2020—WHO Issues Global Health Emergency: With a worldwide death toll of more than 200 and an exponential jump to more than 9800 cases, the WHO finally declares a public health emergency, for just the sixth time. Human-to-human transmission is quickly spreading and can now be found in the United States, Germany, Japan, Vietnam, and Taiwan.

February 3, 2020—US Declares Public Health Emergency: The Trump administration declares a public health emergency due to the coronavirus outbreak. The announcement comes 3 days after WHO declared a Global Health Emergency as more than 9800 cases of the virus and more than 200 deaths had been confirmed worldwide.

The 200 deaths and 9800 cases formed the basis of a global health emergency resulting in a general lockdown of societies worldwide. Even more astonishing is that testing for the specific illness and its stated cause (the virus) did not exist (explained later). The hastily developed test was flawed and had to be abandoned later. However, the story of the virus and its pandemic continued, emphasizing that every claim has a scientific basis.

It was declared to spread, potentially killing millions of people based on computer modeling using purely arbitrary assumptions, i.e., epidemiological and testing.

Recall the previous discussion that there is no definition of "product quality." Still, a colossal bureaucracy at the regulatory authorities (such as the FDA) engages in providing claimed quality products to the public.

This practice of "science" is now extended to the sister branch of bureaucracy called the CDC. It provides no specific definition of the disease and its measurable parameter. Still, it moved on by shutting the population to control the illness by advising the chemical manufacturer to produce chemical compounds (vaccines and pharmaceuticals).

As virologists started the rumor, the new disease had to be virus-based.

The disease was classified as SARS (Severe Acute Respiratory Syndrome) with the associated virus named SAR-CoV-2. In addition, it was promoted that it is a contagious disease and spreads very fast—no scientific or experimental evidence was provided. No virus sample was found or isolated at this stage, and no test was available specific to the virus. Therefore, the idea of the virus and its pandemic has to be imaginary or fictitious, certainly not scientific.

There have been several other claims made in this regard without merit, such as (1) SARS-CoV-2 is contagious; (2) SARS-CoV-2 is 5 or 10 times deadlier than the common flu virus; (3) face masks protect from the virus; (4) social distancing protects the public by stopping or reducing the spread of the virus; (5) washing hands or exposed skin surfaces provides protection from the virus; (6) lockdowns (partial or complete) help reduce the spread of the virus; (7) a significant increase in positive test results (cases) shows a wide spread of the SARS-CoV-2 virus.

However, they all will require an actual physical sample of the virus to support the claims, which is unavailable. Everything is based on self-promoted arbitrary standards and procedures.[157]

It is worth noting that during the past 40 years, not a single study has shown that masks are effective against the virus. Everything is based on a swab test (declared scientific and valid) but without any truth, as explained later.

The authenticity of the claims, being scientific, is based on a peer-review system. Unfortunately, as peers, reviewers, and experts have the same learning experiences, practices, and mindsets, they cannot critically evaluate the lack of science. Hence errors can not be observed and corrected. From the above discussion, it is evident that the medical profession does not follow scientific principles in researching for discovering the virus and its monitoring and developing of treatment for the illness.

As per the definitions provided above, it should be evident that the medical profession has never been trained to conduct such scientific studies. Conducting scientific research is not part of the profession. The profession is making huge errors that damage public health and its professional credibility. Hence such practice should be stopped immediately. For a critical and unbiased assessment, the medical professionals should get their work reviewed by externals to their medical field, particularly chemicals isolation and characterization experts.

Until such time, the medical research practices and publications will remain cult-like presentations and teaching—inaccurate and unscientific. The public will remain observant of nonexistent "new" diseases, treatments, and possibly pandemics.

---

[157] https://www.fda.gov/media/134922/download

Chapter 23—COVID-19 Virus and Pandemic to Vaccines

HERE WE PRESENT a story of deception fueled by not following the science.

There is no doubt that pandemic news has been a worldwide phenomenon. Most countries describe their situation with similar narratives, worries, scares, and the steps for mitigation. The wording may be slightly different, but the end message remains the same.

For example:

> *The COVID-19 pandemic, also known as the coronavirus pandemic, is an ongoing global pandemic of coronavirus disease 2019 (COVID-19) caused by severe acute respiratory syndrome coronavirus 2 (SARS-CoV-2).*[158]

Canadian version:

> *The COVID-19 pandemic in Canada is part of the ongoing worldwide pandemic of coronavirus disease 2019 (COVID-19). It is caused by severe acute respiratory syndrome coronavirus 2 (SARS-CoV-2).*[159]

> *The CDC states that COVID-19 is a respiratory disease caused by SARS-CoV-2, a coronavirus discovered in 2019.*[160]

---

[158] https://en.wikipedia.org/wiki/COVID-19_pandemic

[159] https://en.wikipedia.org/wiki/COVID-19_pandemic_in_Canada

[160] https://www.cdc.gov/dotw/covid-19/index.html

WHO (World Health Organization):

> *COVID-19 is the disease caused by a new coronavirus called SARS-CoV-2. WHO first learned of this new virus on December 31, 2019, following a report of a cluster of cases of 'viral pneumonia' in Wuhan, People's Republic of China.*[161]

UK NHS (National Health Service):

> *SARS (severe acute respiratory syndrome) is caused by the SARS coronavirus, known as SARS CoV. Coronaviruses commonly cause infections in both humans and animals.*[162]

Scientific Journals:

Lancet:

> *A recent cluster of pneumonia cases in Wuhan, China, was caused by a novel betacoronavirus, the 2019 novel coronavirus (2019-nCoV)." "All patients with suspected 2019-nCoV were admitted to a designated hospital in Wuhan. We prospectively collected and analysed data on patients with laboratory-confirmed 2019-nCoV infection by real-time RT-PCR and next-generation sequencing.*[163]

---

[161] https://www.who.int/emergencies/diseases/novel-coronavirus-2019/question-and-answers-hub/q-a-detail/coronavirus-disease-covid-19
[162] https://www.nhs.uk/conditions/sars/
[163] https://www.thelancet.com/journals/lancet/article/PIIS0140-6736(20)30183-5/fulltext

Science:

> *As confirmed cases of a novel virus surge around the world with worrisome speed, all eyes have so far focused on a seafood market in Wuhan, China, as the origin of the outbreak.*[164]

The common denominator is a disease (pneumonia) because of the virus (called SARS-CoV-2 or 2019-nCoV). The news and claim came from the world of medicines (healthcare), particularly regulatory authorities.

So naturally, the public believed as they strongly believed in their health providers and regulatory authorities.

[164] https://www.science.org/content/article/wuhan-seafood-market-may-not-be-source-novel-virus-spreading-globally

Chapter 24—Novel Coronavirus Case Studies

LET US CONSIDER what clinical and scientific evidence provided supporting the start of COVID-19. Below are a few case descriptions of sickness, its severity and the number of deaths worldwide.

Case-1, Report (from the USA, *First Case of 2019 Novel Coronavirus in the United States*, published in The New England Journal of Medicine.[165]

Summary: A 35-year-old man recently returned from China presented to a clinic in Washington State with a 4-day history of cough and subjective fever. The patient visited the hospital voluntarily after hearing of a health alert over an alleged novel coronavirus outbreak in China, apparently with similar symptoms to his own.

The medical examination indicated: body temperature (37.2°C); blood pressure 134/87 mm Hg; oxygen saturation 96%; chest radiography showed no abnormalities. CDC staff decided to test the patient for 2019-nCoV (subsequently renamed SARS-CoV-2) based on current CDC "persons under investigation" case definitions. The patient was discharged from the hospital but was later called back as his PCR test returned a positive reading.

Treatment during this time was largely supportive. The patient received, as needed, 650 mg of acetaminophen every 4 hours and 600 mg of ibuprofen every 6 hours. He also received 600 mg of guaifenesin for his continued cough and approximately 6 liters of normal saline over the first six days of hospitalization.

In view of the potential for hospital-acquired pneumonia,

[165] https://www.nejm.org/doi/full/10.1056/NEJMoa2001191

treatment with vancomycin (a 1750-mg loading dose followed by 1g administered intravenously every 8 hours) and cefepime (administered intravenously every 8 hours) was begun.

The CDC asserted that the patient's nasopharyngeal and oropharyngeal swabs tested positive for SARS-CoV-2 via rRT-PCR assay. The stool and both respiratory specimens also tested positive by rRT-PCR for SARS-CoV-2, whereas the serum remained negative.

Treatment with intravenous remdesivir (a novel nucleotide analog prodrug) was begun on the evening of day 7, and no adverse events were observed in association with the infusion. Vancomycin was discontinued on the evening of day 7, and cefepime was discontinued the following day.

On hospital day 8 (illness day 12), the patient's clinical condition improved. Supplemental oxygen was discontinued, and his oxygen saturation values improved to 94-96% while he was breathing ambient air. The previous bilateral lower-lobe rales were no longer present. His appetite improved, and he was asymptomatic despite intermittent dry cough and rhinorrhea. He was afebrile, and all symptoms had resolved except his cough, which was decreasing in severity.

It is unclear why the patient was subjected to an rRT-PCR test for SARS-CoV-2 when he exhibited what would normally be diagnosed as the mild flu and would therefore be treatable with antibiotics. Nevertheless, because the rRT-PCR test came out positive, the assumption that the patient had SARS-CoV-2 prevailed.

A single-patient observation with otherwise regular flu or pneumonia symptoms and was considered treated successfully formed the basis of the arrival of the deadly disease comparable to the 1918 influenza pandemic. How logical and scientific is that?

Case-2, Report from Australia, *Isolation and Rapid Sharing of the 2019 Novel Coronavirus (SARS-CoV-2) from the First Patient*

Diagnosed with COVID-19 in Australia[166]

Summary: A 58-year-old man from Wuhan, China, felt unwell on the day of his arrival in Melbourne (19 January 2020). In China, he had had no contact with live food markets, people known to have COVID-19, or hospitals. He developed a fever on 20 January and a cough with sputum production; on 24 January, he was admitted to the Monash Medical Centre, Melbourne, from its emergency department with progressive dyspnoea. His temperature was 38.1°C, his heart rate 95 beats/min, and O_2 saturation 94% on room air. A chest x-ray showed subtle ill-defined opacities.

Intravenous ceftriaxone (2 g/day) and azithromycin (500 mg/day) were commenced on admission day 4 to treat potential secondary bacterial pneumonia, although no bacterial pathogen was identified. Low-flow oxygen (maximum 3 L/min via nasal prongs) was administered until admission day 10. The patient gradually improved; fever, productive cough, and dyspnoea resolved by admission day 12, and he was discharged from the hospital on 7 February (admission day 15).

It is unclear why the patient was subjected to an rRT-PCR test for SARS-CoV-2 when he exhibited what would normally be diagnosed as the mild flu and would therefore be treatable with antibiotics.

Case-3, Report from Wuhan China, *A Novel Coronavirus from Patients with Pneumonia in China,* 2019, The New England Journal of Medicine[167]

Summary: Three adult patients presented with severe pneumonia and were admitted to a hospital in Wuhan on December 27, 2019. Patient 1 was a 49-year-old woman, Patient 2 was a 61-year-old man, and Patient 3 was a 32-year-old man. Clinical profiles were

[166] https://onlinelibrary.wiley.com/doi/full/10.5694/mja2.50569

[167] https://www.nejm.org/doi/full/10.1056/nejmoa2001017

available for Patients 1 and 2. Patient 1 reported having no underlying chronic medical conditions but reported fever (temperature, 37°C to 38°C) and cough with chest discomfort on December 23, 2019. Four days after the onset of the illness, her cough and chest discomfort worsened, but the fever was reduced; a diagnosis of pneumonia was based on a computed tomographic (CT) scan. Patient 2 initially reported fever and cough on December 20, 2019; respiratory distress developed seven days after the onset of illness and worsened over the next two days, at which time mechanical ventilation was started. Patients 1 and 3 recovered and were discharged from the hospital on January 16, 2020. Patient 2 died [via mechanical ventilation?] on January 9, 2020. No biopsy specimens were obtained.

Case-4, Report from South Korea, *The First Case of 2019 Novel Coronavirus Pneumonia Imported into Korea from Wuhan, China: Implication for Infection Prevention and Control Measures*[168]

Summary: A 35-year-old woman who lives in Wuhan, China, arrived at the Incheon Airport (South Korea) on January 19, 2020. During the quarantine inspection process at the airport, her body temperature was 38.3°C by a thermal scanner; therefore, she was immediately hospitalized in a designated isolation hospital. Coronavirus conventional polymerase chain reaction (PCR) assay was positive for the throat swab sample, and sequencing of the PCR amplicon showed that the sequence was identical to that of the 2019-nCoV isolated from the Wuhan patient.

One day before entry (January 18, 2020), she developed fever, chill, and myalgia and visited a local Wuhan clinic. Chest radiography showed no infiltrates, and she was diagnosed as a common cold. On admission (January 19, 2020), the physical examination revealed a body temperature of 38.4°C, a respiratory

---

168 *J Korean Med Sci. 2020 Feb 10;35(5):e61,* https://jkms.org/DOIx.php?id=10.3346/jkms.2020.35.e61

rate of 22 breaths per minute, a pulse of 118 per minute, and blood pressure of 139/92 mmHg. The initial chest radiography showed no infiltrations. The laboratory tests showed mild changes. She developed nasal congestion, cough, sputum, pleuritic chest discomfort, and watery diarrhea during admission.

On January 21, 2020 (day 4 of illness), she did not complain of shortness of breath, but oxygen supplementation via nasal cannula (3 L/min) was started as a result of decreased arterial oxygen saturation of around 91%. Her oxygen requirement increased to 6 L/min on January 24, 2020 (day 7 of illness), and chest radiography taken on January 25, 2020 (day 8 of illness) began to show chest infiltrates in the right lower lung field.

After the diagnosis of 2019-nCoV infection was made, lopinavir 400 mg/Ritonavir 100 mg was given from January 21, 2020 (day 4 of illness). The fever persisted for ten days, with a maximum temperature of 38.9°C on day 7 of illness, and then subsided on January 28, 2019 (day 11 of illness). From January 31, 2019 (day 14 of illness), her dyspnea began to improve, reducing oxygen requirement, and the lung lesions also began to diminish in chest radiography. The only clue to 2019-nCoV infection was her travel history.

Case 5, Report from Canada, Toronto, *Canada's first presumptive case of the novel coronavirus has been officially confirmed, Ontario health officials said Monday as they announced the patient's wife has also contracted the illness.*[169]

Summary: Dr. Barbara Yaffe, the province's associate chief medical officer of health, said the confirmation came through earlier in the day following tests at the National Microbiology Laboratory in Winnipeg. A Toronto man in his mid-50s had initially tested positive at a provincial facility days after returning to Toronto from

[169] https://nationalpost.com/news/canada/ontario-confirms-second-coronavirus-case-the-wife-of-the-first-case

Wuhan—the virus's epicenter in China—via Guangzhou.

Case 6, Report from Germany, *Complete Genome Sequence of a SARS-CoV-2 Strain Isolated in Northern Germany.*[170]

Summary: The isolate was obtained from an upper respiratory tract specimen (oropharyngeal swab) from a 62-year-old woman who was part of a small local cluster of COVID-19 cases, originating from a household contact who likely acquired the virus while traveling in Italy. The patient herself did not show any signs or symptoms at the time the sample was taken. [3]

The swab sample (ESwab; Copan, Italy) tested SARS-CoV-2 RNA positive at the Institute for Microbiology, Virology, and Hygiene at the University Medical Center Hamburg-Eppendorf by real-time reverse transcription-PCR (RT-PCR), as described previously. [4] For virus isolation, 500 µl of the remaining (not inactivated) ESwab medium was used for adsorption on Vero cells (ATCC CRL-1586) seeded in T25 flasks.

After 1 h at 37°C, the ESwab medium was replaced by cell culture medium (Dulbecco's modified Eagle's medium containing 3% fetal calf serum, 1% penicillin, streptomycin, 1% L-glutamine [200 mM], 1% sodium pyruvate, and 1% nonessential amino acids [all from Gibco/Thermo Fisher Scientific, Waltham, MA, USA]).

Cells were monitored daily for cytopathic effect (CPE). No CPE was observed until day 4, but virus growth was confirmed by real-time RT-PCR. Additional experiments revealed that infection of different Vero cells (ATCC CCL-81) resulted in a strong CPE at 24 to 48 h postinfection.

For sequencing of the viral RNA, nucleic acid extraction was performed automatically using the QIAsymphony DSP virus/pathogen kit and 200µl cell culture supernatant (supplemented 1:1 with Roche PCR medium for inactivation).

[170] https://journals.asm.org/doi/10.1128/MRA.00520-20

Chapter 25—Case Study Summary

TO SUMMARIZE, IT is clear that the patients appear to have flu or pneumonia symptoms or illnesses treatable with traditional therapies. The illness (pandemic) observations were mainly based on single patient observations (scientific, statistical, etc.?). Most patients made successful recovery except one in China who died, for which the cause is uncertain.

Compared to the 1918 influenza pandemic, there was no evidence that the illness was deadly and a virus or disease. Perhaps the most critical aspect is that all decisions about the disease and pandemic were based on the PCR test, which, as explained earlier, is neither a valid test for virus nor illness. Some exaggerated claims of disease, spread, and severity of the unknown disease were made, such as:

> *...the novel strain of coronavirus causing the devastating global COVID-19 pandemic.*[171]

> *The COVID-19 pandemic is now a major global health threat. As of March 16, 2020, there have been 164,837 cases and 6,470 deaths confirmed worldwide. The global spread has been rapid, with 146 countries now having reported at least one case. The last time the world responded to a global emerging disease epidemic of the scale of*

[171] Claimed Stephen Hoge, the President of Moderna, Inc., at the Hearing Before the House Energy and Commerce Committee Subcommittee on Oversight & Investigations (July 21, 2020). https://energycommerce.house.gov/sites/democrats.energycommerce.house.gov/files/documents/Testimony%20-%20Hoge%2020200721_0.pdf

> *the current COVID-19 pandemic with no access to vaccines was the 1918-19 H1N1 influenza pandemic.*[172]

From reporting of the first case, "*HOW first learned of this new virus on December 31, 2019*" to March 16, 2020, for a write-up of the report (i.e., within two and half months), the claimed virus has been isolated, identified, characterized, linked to the disease with a prediction of its severity comparable to 1918-19 pandemic.

Wow! Science? Sorry, it is CDC science—word of mouth—cannot be science or actual science.

The prediction of deaths in the GB and US was 510,000 and 2,200,000, respectively, within six months.

Being a scientist all my life, including working at Health Canada for 30 years in the area of pharmaceuticals/medicines evaluation, I (Saeed Qureshi) am well aware of the lack of competencies at the authorities, including the FDA and other regulator agencies, because of close interaction with my counterpart there. I have published my work in scientific journals and written on my blogs about the funny prevailing science at the authorities.[173] [174]

For example, US FDA scientists and scientists from worldwide agencies claim they establish the quality (safety and efficacy) of consumers'/patients' products. This claim is scientifically invalid and a lie. The Agency neither defines a quality product nor provides a valid test method to establish it, as explained in the scientific part earlier. Instead, the FDA works under its own version of science called "regulatory science," described as:

> *Regulatory science encompasses a wide range of*

---

[172] Ferguson et al, https://www.imperial.ac.uk/media/imperial-college/medicine/mrc-gida/2020-03-16-COVID19-Report-9.pdf

[173] www.bioanalyticx.com

[174] www.drug-dissolution-testing.com

> *subjects, including not only disciplines traditionally associated with regulation, such as statistics and clinical research, but also disciplines outside the biomedical sciences such as economics, risk communication, and sociology.*[175]

Note that there is no mention of actual science (chemistry) on which the practice of medicines is based. All the medical and pharmaceutical literature details chemical descriptions, as explained earlier.

FDA website further elaborated its version of science called Regulatory Science. "Regulatory Science is the science of developing new tools, standards, and approaches to assess the safety, efficacy, quality, and performance of all FDA-regulated products." [176]

With this background, I became suspicious about the claims made by CDC, and indeed it became evident that the virus and the pandemic have no scientific basis. I described such shortcomings through my blog as most scientific avenues (journals, publications, etc.) would not consider such views. They are not in line with commonly accepted FDA or CDC science narratives—and by peers in the field of medicine.

The rest of the section deciphers the confusion, in fact, the lies, in more detail, presented during the past two years, logically while considering the science background described earlier in the chapter. It is important to study and seek help from the first part to understand the confusing jargon often used in the medical and pharmaceutical literature.

Let us follow the pandemic narrative. First, a disease called

---

[175] https://www.ncbi.nlm.nih.gov/books/NBK92887/

[176] https://www.fda.gov/science-research/science-and-research-special-topics/advancing-regulatory-science#:~:text=Regulatory%20Science%20is%20the%20science,of%20all%20FDA%2Dregulated%20products

COVID-19 is caused by a virus called SARS-CoV-2, which can be tested with a PCR test. The disease is characterized by symptoms noted on the CDC website.[177]

> *People with COVID-19 have had a wide range of symptoms reported—ranging from mild symptoms to severe illness. Symptoms may appear 2-14 days after exposure to the virus. However, anyone can have mild to severe symptoms. People with these symptoms may have COVID-19:*
>
> *Fever or chills, cough, shortness of breath or difficulty breathing, fatigue, muscle or body aches, headache, new loss of taste or smell, sore throat, congestion or runny nose, nausea or vomiting, diarrhea.*

Unfortunately, there is no newness or uniqueness to these symptoms. These are common flu-like symptoms. Everyone experiences such symptoms once in a while, particularly when the season changes.

Symptoms hardly ever indicate or establish a specific illness. Instead, the illness is established and confirmed usually by a test. The test mentioned in this case is the PCR (Polymerase Chain Reaction) test.

The first lie, based on ignorance of science, is that it is a virus test—it is not. The second lie is that it is a test for illness or COVID-19. It is not. The test does not detect the virus, disease, or infection. However, it is described and endorsed for the virus and illnesses under the deceitful label of cases.

It is an RNA test. An RNA is a chemical compound presumed to be part of the virus. So, the best one could know from the test is that it detects RNA, not the virus, if it is positive. However, the

---

[177] https://www.cdc.gov/coronavirus/2019-ncov/symptoms-testing/symptoms.html

test is conducted using a small portion of solution after treating the swab sample with multiple chemicals (30+) and adding cell culture of African green monkey kidney. Therefore, the test lacks evidence that RNA belongs to the virus.

Next, a PCR test requires a chemical compound called primer for reaction (test) to create and increase/multiply the specific RNA one is looking for. It means the test does not test or see; the analyst decides what to produce and see. Therefore, one cannot determine new or novel (unknown) RNA (labeled as a virus). Instead, by adding a primer of one's choice, one will obtain an expected RNA or its variant, not a new one, of one's choice or liking.

The question is: why is such silliness happening? The reason, the test, its development is part of science subject, in particular chemistry. However, it is developed by medical and pharmaceutical experts, particularly virologists who have never studied or trained in developing a test.

It is as if medical experts develop a sound-based test/tester for evaluating the performance of car engines because they have been driving cars for decades and know the sounds of well-performing and non-performing car engines. This is not a joke, but it describes the approach medical experts, virologists, immunologists, and microbiologists use to develop a PCR test for a virus or its variant.

The test was being developed while the virus was declared to exist and spread. How could one detect or know that a virus (new or old) is there when a test is unavailable? Could one find gold ores or mines when one does not have a well-established test or procedure to measure the gold content?

Of course not, but medical professionals (virologists in particular) magically can.

Chapter 26—Too Many Coincidences for It Not to Be a Plandemic

FOR SPEAKING OUT from her experience as an insider on the vaccine industry, American virologist Dr Judy Mikovits, has been vilified and de-platformed about as much as any independent scientist whistleblower.

Mikovits has been persuasive in her dissection of the malfeasance behind COVID-19. Her heroic battle, to expose the truth about the true machinations hidden from public view, show us that this was all about; control, obscene profits, corrupting individuals and institutions, and using public health as a weapon against the masses.

Like our own STEM researchers at Principia Scientific International, Dr Mikovits identifies a core of corrupt individuals who are the architects of what any rational mind would conclude was a highly coordinated, planned eugenicist war on humanity.

The ill-informed, of course, scoff at the mere suggestion that an elite cohort of carefully selected individuals would have the means or the willing cooperation of so many diverse international bodies, esteemed scientific and academic institutions, disparate business interests and be able to evade detection by keen-eyed international journalists eager to scoop such an incredible story.

For instance, while GDP in most nations is taking a nosedive to minus 33 percent, billionaires grew their wealth at rates not seen since the heady days of the railroad tycoons over a century ago.

An article in The Washington Post detailed some of the divides.[178]

178

https://www.washingtonpost.com/business/2021/04/06/billionaire-wealth-forbes-pandemic/

Thanks to the lockdown measures, the number of billionaires on Forbes' 35th annual ranking swelled by 660 to 2,755—a roughly 30 percent jump from the year before—and 493 of them are first-timers. Amazon seized 42 percent of the e-commerce as the pandemic fueled a surge in online shopping. As a class, billionaires added about $8 trillion to their total net worth from the previous year, totaling $13.1 trillion.

The Bill and Melinda Gates Foundation, meanwhile, has churned out announcements and articles, including their latest one, *Why We're Giving $250 Million More to Fight Covid-19.* The Foundation boasts that it has given a total of $1.75 billion to pandemic efforts.

At the time of writing, the Gates Foundation made its largest single contribution to fight the pandemic—$250 million. Why so much? And why do so roughly a year after COVID-19 first appeared?

The reality is that every meaningful aspect of our society is controlled by an elite that has invested generations in seizing and maintaining control for their own ends.

Scratch the surface of the 'great and the good' prominent in the public eye in managing governmental response to the pandemic and you will unearth a connected matrix of coincidences.

With a serious examination of individuals, such as Dr Anthony Fauci and Bill Gates, and organizations like the CDC, NIH, WHO, and Bill & Melinda Gates Foundation, among others, there is a dark force uniting them all and driving the global vaccination agenda.

Bill Gates, no less, felt compelled to admit in December 2021 that vaccines do not stop the spread of COVID.

On Twitter he wrote:

> *We didn't have vaccines that block transmission. We got vaccines that help you with your health but they only slightly reduce transmission. We*

> *need a new way of doing the vaccines.*[179]

But his Twitter admission was not reported widely. Ironically, an important study published on medRxiv, which was a joint project of Cold Spring Harbor Laboratory, Yale University and the British Medical Journal:

> *...found no statistically significant difference in transmission potential between vaccinated persons and persons who were not fully vaccinated.*[180]

At the same time Gates was admitting the vaccines were less than perfect than as advertised. Cornell University, with a 100% vaccination rate, shut down due to an outbreak of COVID. Of course, such important studies are swept under the carpet by the media when they run counter to mainstream fear porn.

Books like our own are a threat to their hegemony. Our authorship credentials are the several decades we have spent working within this rotten system.

To us, the brainwashing of immense numbers of the population is staggering to behold. Our adversaries are the mainstream's 'fact checkers' who have recently been outed in litigation as nothing more than biased opinion peddlers.

As reported in the New York Post (December 14, 2021) a lawsuit brought by celebrated journalist John Stossel, has exposed the mainstream media's supposed battle against misinformation as a farce.[181] When independent journalist, Stossel, took Facebook to court for defaming him as a lying conspiracy theorist the social

---

[179] https://twitter.com/MaajidNawaz/status/1471218073214541835

[180] https://www.medrxiv.org/content/10.1101/2021.11.12.21265796v1.full.pdf

[181] https://nypost.com/2021/12/14/facebook-admits-the-truth-fact-checks-are-really-just-lefty-opinion/

media giant's lawyers conceded in their court papers that Facebook's fact-checks are merely opinion.

It has also transpired that in April 2020, Facebook CEO, Mark Zuckerberg, offered National Institute of Allergy and Infectious Diseases (NIAID) Director Anthony Fauci help in facilitating decisions regarding lockdown measures in the US, as revealed by private emails exchanged between the two.

According to The National Pulse, Zuckerberg's email offer aggregating anonymized user reports that would help NIAID decide whether to tighten or loosen lockdown mandates.[182] Some emails show them discussing a coronavirus information hub being set up on the giant social media site as a means to steer the messaging in the desired direction.

The emails allegedly show Zuckerberg willing to put Facebook's gigantic user data trove at NIAID's disposal in user reports form and came a month after the pair started communicating directly—a revelation stemming from another batch of government-redacted emails Zuckerberg and Fauci sent to each other in March 2020.[183]

Zuckerberg also expressed willingness to provide more *resources* in order to speed up vaccine development. In his response, Fauci said, *"I will think hard about ways that we may take you up on your offer."*

So, the connivance is real.

But so blatant is the media bias in favor of the Covidian cult and so unabashed in quelling any attempt by medical professionals to expose the criminality, even the British Medical Journal (BMJ), not known for entering into conspiracy theorisation, published the following open letter condemning Facebook:

---

[182] https://www.thethinkingconservative.com/new-fauci-zuckerberg-emails-reveal-offer-of-data-reports-to-aid-lockdown-policies-vaccine-development/

[183] https://reclaimthenet.org/zuckerberg-offered-user-data-reports-to-fauci-to-help-shape-lockdown-policies/

Rapid Response:
Open letter from The BMJ to Mark Zuckerberg
Dear Mark Zuckerberg,

We are Fiona Godlee and Kamran Abbasi, editors of The BMJ, one of the world's oldest and most influential general medical journals. We are writing to raise serious concerns about the "*fact checking*" being undertaken by third party providers on behalf of Facebook/Meta.

In September, a former employee of Ventavia, a contract research company helping carry out the main Pfizer COVID-19 vaccine trial, began providing The BMJ with dozens of internal company documents, photos, audio recordings, and emails. These materials revealed a host of poor clinical trial research practices occurring at Ventavia that could impact data integrity and patient safety. We also discovered that, despite receiving a direct complaint about these problems over a year ago, the FDA did not inspect Ventavia's trial sites.

The BMJ commissioned an investigative reporter to write up the story for our journal. The article was published on 2 November, following legal review, external peer review and subject to The BMJ's usual high level editorial oversight and review. [1]

But from November 10, readers began reporting a variety of problems when trying to share our article. Some reported being unable to share it. Many others reported having their posts flagged with a warning about "*Missing context...Independent fact-checkers say this information could mislead people.*"

Those trying to post the article were informed by Facebook that people who repeatedly share false information might have their posts moved lower in Facebook's News Feed. Group administrators where the article was shared received messages from Facebook informing them that such posts were partly false.

Readers were directed to a fact check performed by a Facebook contractor named Lead Stories. [2]

We find the "fact check" performed by Lead Stories to be inaccurate, incompetent and irresponsible.

Fiona Godlee, editor in chief
Kamran Abbasi, incoming editor in chief
The BMJ

The Great Barrington Declaration was not an outright challenge to the authorities but was a measured and diplomatic effort to win hearts and minds over to a moderate path of acceptance that the pandemic was real but that only those most at risk should be targeted for intervention.

By contrast, the World Council for Health (WCH), a worldwide coalition of health-focused organizations and civil society groups, declared that COVID-19 "vaccines" are dangerous and unsafe for human use. The WCH pulled no punches in an outright declaration that the manufacturing, distribution, administration, and promotion of these injections violate basic law principles. The Council further declared that any direct or indirect involvement in the manufacturing, distribution, administration and promotion of COVID jabs violates basic principles of common law, constitutional law and natural justice, as well as the Nuremberg Code, the Helsinki Declaration, and other international treaties.[184]

[184] https://worldcouncilforhealth.org/campaign/covid-19-vaccine-cease-and-desist/

Chapter 27—The Big Picture: Going for the Jugular

ANDREAS OEHLER HAS done a sterling job in setting out how the World Economic Forum seems to be the driver behind and organizer of the global population control operations with his analysis.[185]

He identifies that the pandemic is the cleverly crafted cover for the Great Reset in which fascist corporatism will usher in biometric IDs, a cashless society, universal basic income (UBI) all in the name of the common good.

This is also a eugenicists dream scenario, because at the root of the billionaire class is their hatred of the common man and woman and nothing will end well when we buy into the concept that the "common good" means we sacrifice our individual sovereignty to the psychopathic architects of it all.

First, the lieutenants trained to lead the lower ranks were schooled within the World Economic Forum (WEF). The field commanders are within the Bill and Melinda Gates Foundation (BMGF), a sponsor of the prescient "*Event 201.*"[186]

We learn that it is the:

> *...coronavirus pandemic simulation exercise, held in New York City on October 18, 2019—the same day as the opening of the Wuhan Military World Games, seen by some as "ground zero" of the global pandemic.*

Andreas Oehler was rightly intrigued to find that there was a bizarre coincidence here in that the original release of the James

---

185 https://live2fightanotherday.substack.com/p/going-for-jugular

186 https://www.centerforhealthsecurity.org/event201/

Bond film *No Time to Die* was scheduled for October 25, 2019, but this was later switched.

In the movie, millions of indestructible DNA-aware nanorobots capable of spreading through touch from human to human are unleashed.

Oehler explains:

> *Upon encountering its intended target though a DNA match, the weapon could kill within seconds. However, with alterations the technology could become a weapon of mass destruction—capable of targeting everything from individuals to entire ethnicities. Are the plot and timing coincidental?*

Not a completely silly question, if you look at all such coincidences and the picture they paint.

Oehler then tells us that the WEF has been a leading proponent of digital biometric identity systems, arguing that they will make societies and industries more efficient, more productive and more secure. In July 2019, the WEF started a project to *"shape the future of travel with biometric-enabled digital traveler identity management."*

But the WEF is but one player in the game and there are others, mostly faceless and very unlikely to be known to anyone who does not make it their business to go deep down the rabbit hole.

Yes, the project is managed by the World Economic Forum in Davos, and openly advertised under various names: the Fourth Industrial Revolution, the Great Reset, and Agenda 2030.

The Fourth Industrial Revolution is a term readily adopted in mainstream schools as a catchphrase for a new Utopia that will likely result in our further enslavement but teaching union bureaucrats are pushing it to please their elite puppet masters.

The upcoming technological reset is more like a dystopian

science fiction novel with the iconic title, the Fourth Industrial Revolution—or 4-IR for short. The first, second, and third industrial revolutions were the mechanical, electrical, and digital revolutions. The 4-IR marks the convergence of existing and emerging fields, including Big Data, artificial intelligence, machine learning, quantum computing, genetics, nanotechnology, and robotics. What it will usher is the merging of the physical, digital, and biological worlds. This will challenge the ontologies by which we understand ourselves and the world, including the definition of a human being. In short, it is transhumanism made real. Few people realise that a mass consolidation of nearly all the world's teachers' unions under a single Global Union Federation (GUF) has been taking shape this century and which was serve to indoctrinate the young to accept this all as humanity's inevitable future. It is all part of a WEF and UNESCO master plan through an entity known as Education International (EI).[187]

EI combines 383 member organisations including America's AFT and the NEA, and collaborates with the WEF[188] and UNESCO.[189]

EI has been galvanizing teaching unions (and thus, all teachers) into conformity with the "reimagine" ed-tech agendas of these global governance institutions. To put it bluntly, EI puppeteers the AFT and the NEA using marionette strings that are tied to the WEF and UNESCO. In turn, the AFT and the NEA, along with the other hundreds of member organizations belonging to EI, are being spurred to "reimagine" schools through corporate ed-tech innovations geared towards advancing the Fourth Industrial Revolution[190] (4IR) that is being accelerated by the WEF and the

---

[187] https://www.ei-ie.org/en/about/who-we-are

[188] https://acc2021.ei-ie.org/en/item/20483:mondial

[189] https://www.ei-ie.org/en/item/25127:education-international-the-oecd-and-unesco-j

[190] https://www.ei-ie.org/en/item/20086:the-importance-of-teachers-and-the-so-called-fourth-industrial-revolution

United Nations (UN) through the policy agendas collectively known as the Great Reset. No surprise to skeptics of man-made global warming that the avid eugenicist and climate alarmists, Al Gore, is a key promoter.[191] Heed the words of Malcolm X:

Only a fool would let his enemy teach his children.

Gore and his ilk are pushing this agenda with the narrative it will stem the next great crisis the world will face when the COVID-19 pandemic finally subsides climate change.

Thus, we go full circle from one Chicken Little escapade to another, with always the same psychopaths at the helm of key institutions.

Those bodies more readily linked to the conspiracy include the Rockefeller Foundation, the Bill and Melinda Gates Foundation, the Wellcome Trust, the World Health Organisation, the Chinese Communist Party and BlackRock. It is no coincidence these same people are the self-appointed Coalition for Epidemic Preparedness Innovations about which Paula Jardine wrote of in *Defending Freedom* at conservativewoman.co.uk.[192]

The Gates and Rockefeller foundations run their program to "provide digital ID with vaccines" and they call the vaccination of children "an entry point for digital identity."

In truth, this means that everything from employment, travel, commerce, health care and more will require every individual to have a digital ID. It means we will no longer have the right to privacy and the authorities administering the system will have the power to remove you from any activity you ordinarily expect to be able to access.

This is all sold to us under the guise of *Build Back Better* and

---

[191] https://www.foxbusiness.com/markets/al-gore-un-secretary-general-great-reset-global-capitalism

[192] https://www.conservativewoman.co.uk/forget-china-it-was-americas-bio-spooks-who-locked-down-the-west/

the WEF is strident in positioning itself as a global leader in the Great Reset hopium that would dupe many docile souls into believing some good must come of it all. It won't.

We know this if we take to read what WEF founder Klaus Schwab tells us in his book *COVID-19: The Great Reset.* Schwab and his psychopathic cronies want you to swallow the lie that the pandemic was an unexpected blessing that can be utilized to usher in a transformed (transhumanist) world.

They want global government overseen by the ultra-rich who will oversee the new "economic, societal, geopolitical, environmental and technological reset."

As with man-made global warming—the playbook is the same as are the characters behind the curtains. Global warming took hold in the international political arena in July 1988 when UK Prime Minister, Margaret Thatcher, gave her pivotal speech to Royal Society, London about the impending threat of greenhouse gases. Shortly thereafter, NASA's Dr James Hansen gave his own doomsaying speech in Washington DC to trigger America's path to cutting carbon dioxide (CO_2) emissions.

Those of us who have fought for decades to expose the junk science lies about 'heat trapping' CO_2 were initially caught unawares, in October 2020, when the dreaded COVID-19 pandemic reared its ugly (fake) head. We had been blindsided and it took many months to discern that the same players, many via the undemocratic private club, United Nations, were running both scams.

We learned that the WEF was set up in 1993 as a program called *Global Leaders for Tomorrow.* It would be rebranded, in 2004, as *Young Global Leaders,* a program geared towards "identifying, selecting and promoting future global leaders in both business and politics." [193]

By stealth what was put into place over many years were compliant political leaders who were groomed to do the bidding of

---

[193] https://www.younggloballeaders.org/

the uber-rich eugenicists.

Indeed, as Andreas Oehler correctly identified, quite a few *Young Global Leaders* have later managed to become Presidents, Prime Ministers, or CEOs. The list of the chosen included:

- Microsoft founder Bill Gates (1993),
- California Governor Gavin Newsom (selected in 2005),
- Pete Buttigieg (selected in 2019, candidate for US President in 2020, US secretary of transportation since 2021),
- Stéphane Bancel (Moderna CEO),
- Facebook founder and CEO Mark Zuckerberg (2009),
- Facebook COO Sheryl Sandberg (2007),
- Google co-founders Sergey Brin and Larry Page (2002/2005),
- Covid Twitter personality Eric Feigl-Ding (a 'WEF Global Shaper' since 2013),
- New Zealand Prime Minister Jacinda Ardern (since 2017, selected in 2014),
- Australian Health Minister Greg Hunt (selected in 2003; former WEF strategy director)
- Canadian Deputy Prime Minister Chrystia Freeland (selected in 2001; former managing director of Reuters),
- Canadian Prime Minister Justin Trudeau is a WEF participant, but is not a confirmed Young Global Leader,
- German Chancellor Angela Merkel (selected in 1993, 12 years before becoming Chancellor),
- Current German Health Minister Jens Spahn and former Health Ministers Philipp Roesler and Daniel Bahr,
- EU Commission Presidents Jose Manuel Barroso (2004-2014, selected in 1993) and Jean-Claude Juncker (2014-2019, selected in 1995),
- French President Emanuel Macron (since 2017, selected in 2016),
- Former French President Nicolas Sakozy (2007-2012, selected in 1993),
- Austrian Chancellor Sebastian Kurz,

- Former Italian Prime Minister Matteo Renzi (2014-2016, selected in 2012),
- Former Spanish Prime Minister Jose Maria Aznar (1996-2004, selected in 1993).

There are many familiar actors in the horror story. Not only the WEF cult adherents were recruited from the useful idiots already in the position of power, but many have been groomed and tasked with WEF pet projects for many decades. As Henry Kissinger concluded in his thesis:

> *...a mystic relationship to the Infinite provides the foundation for motivated activity, to which the phenomenal world can offer no permanent obstacles.*

Indeed, job well done, Henry! For those younger reader who have not heard the name, Dr Henry Kissinger, German-born American politician and winner of the Nobel Peace Prize and one of the most evil men alive.

Kissinger was a key figure back during the Richard Nixon administration in bringing China into world economic prominence. Not a philanthropist, despite his Peace Prize, as we can discern from reading his book, *The Final Days.*[194]

Kissinger's book said that "The elderly are useless eaters" and that "Power is the ultimate aphrodisiac." From his National Security Memo number 200 dated April 24, 1974, he wrote:

> *Depopulation should be the highest priority of foreign policy towards the third world, because the US economy will require large and increasing amounts of minerals from abroad, especially from less developed countries.*

---

[194] https://en.wikipedia.org/wiki/The_Final_Days

Just as pertinent to today was Kissinger's speech at Evian, France on May 21, 1992 at a Bilderburg meeting (Kissinger is widely regarded as the architect of most of their machinations), where he eerily said:

> *Today Americans would be outraged if U.N. troops entered Los Angeles to restore order; tomorrow they will be grateful. This is especially true if they were told there was an outside threat from beyond, whether real or promulgated, that threatened our very existence. It is then that all peoples of the world will plead with world leaders to deliver them from this evil. The one thing we all fear is the unknown. When presented with this scenario, individual rights will be willingly relinquished for the guarantee of their wellbeing granted to them by their world government.*

What could be more forewarning of today's invisible threat, that deadly SARS-CoV-2 pathogen?

As Oehler and other analysts have noted, as the pandemic's start date neared Bill Gates of BMGF and Dr Anthony Fauci of NIH acquired control of the WHO. Thereafter, with the help of Swissmedic, governments got into line with GAVI (Global Alliance for Vaccines and Immunization).

Founded in 2000, the GAVI Alliance consisted of the WHO, UNICEF, the World Bank and the Bill and Melinda Gates Foundation.[195]

GAVI is notorious for catastrophic mass vaccination programs for polio in Africa and India. The vaccines were to fight non-existent epidemics. But the vaccine allegedly resulted in far more deaths than the virus and lawsuits against Gates were the outcome.

The tie up this neat nexus of evil Andreas Oehler points to the brilliant work of Dr David Martin who identifies Anthony Fauci in

[195] https://www.gavi.org/

NIAID/NIH as the prime promoter of the gain-of-function research on coronaviruses, S spikes and mRNA "vaccines" (through Moderna) throughout 1998-2019.[196]

The Moderna story as revealed by by Dr David Martin.[197]

- April 2020, Anthony Fauci sat in the President's Oval Office and talked about how Moderna was going to be the mysterious savior of the world, despite the fact that Moderna had never produced a safe commercial product in its entire operating history.
- What you don't remember is that in 2010, it was the 10th anniversary of the funding that actually started Moderna. You don't know what the funding that started Moderna is because nobody talks about it. The funding that started Moderna was the National Science Foundation grant called *Darwinian Chemical Systems.*[198] Darwinian Chemical Systems: In a post-extinction event we want to see if we can get mRNA to write into DNA the code to start human evolution again. That's the 10-year grant that started Moderna.
- Darwinian Chemical Systems was to use RNA to write into the DNA of life. The 20-year funding record of the company that's actually doing it proved that this is what happened.

It is little wonder then that Kissinger, Schwab, Gates and Fauci are increasingly being called the *Four Horsemen of the Apocalypse*.

Dr Martin, so incensed by the inhumanity displayed by these psychopaths had this to say on November 5, 2021, in his presentation at the Wise Traditions Conference:

> *I am going to put my life energy on a single focus. And that is until we hold the criminal conspiracy accountable, and we end the emergency use*

[196] https://en.wikipedia.org/wiki/Moderna

[197] https://richardsonpost.com/harryrichardson/24260/dr-martin-boots-the-elite/

[198] https://www.nsf.gov/awardsearch/showAward?AWD_ID=0434507 https://psuvanguard.com/chemistry-grant-takes-aim-at-lifes-origins/

> *authorization we're not done. And there is no other topic, right now, and I don't care what topic you love, there is no other topic that I am concerning myself with save the preservation of the next generation from a mass murdering, genocidal, psychopathic group of criminals who've decided that they can look at a five-year-old with contempt.*[199]

While Dr Martin and others are doing sterling work in the U.S. to bring to justice the criminal COVID cabal there, in the U.K. concerned citizens and lawyers are similarly pursuing the guilty.

---

[199] https://odysee.com/@tcs1949:4/Dr-David-Martin-speech-at-the-wise-traditions-conference:0d5

Chapter 28—Botched Coronavirus Testing

IT IS A well-established fact that the test was not available at the start of the pandemic, and the hurriedly developed test was faulty and had to be recalled; for example, *Why the CDC botched its coronavirus testing*, MIT Technology Review, stating:

> *Few health institutions around the world are as renowned as the US Centers for Disease Control and Prevention, which makes it all the more baffling that the CDC could have fumbled the rollout of coronavirus diagnostic tests throughout the country so badly.*[200]

A recent report provided a more explicit analysis of the situation, *CDC's First COVID Tests had Design Flaw*, stating:

> *An internal review sheds new light on what went wrong with the first COVID tests distributed by the CDC during the early days of the pandemic.*
>
> *Previous investigations said the contamination was the major reason tests shipped to health labs in early 2020 produced inconclusive reports and false positives.*
>
> *But the CDC's internal review published in Plos O.N.E. says a design flaw also caused problems with the testing kits.*
>
> *The test kits were designed to detect the virus with primers, which bind to and copy*

[200] https://www.technologyreview.com/2020/03/05/905484/why-the-cdc-botched-its-coronavirus-testing

> *targeted sequences, and with probes that emit a fluorescent signal when copies are made, The New York Times reported. The fluorescent signal means the virus's genetic material is present.*
>
> *The probes and primers were not supposed to touch or bind to each other, but that happened sometimes in the faulty kits. And this created the false positives, The New York Times said.*[201]
>
> *By early February 2020, the CDC admitted the tests weren't working and redesigned them with the help of outside laboratories, The New York Times said.*[202] [203]

So, all the claims were based on an admittedly flawed test. Should not the professionals and experts have moral or professional responsibility to admit the flaw and stop the spread of false pandemic news and scare?

---

[201] https://www.nytimes.com/2021/12/15/health/cdc-covid-tests-contaminated.html

[202] https://www.webmd.com/lung/news/20211216/report-says-cdcs-first-covid-tests-had-design-flaw

[203] https://www.webmd.com/lung/news/20211216/report-says-cdcs-first-covid-tests-had-design-flaw

Chapter 29—CDC Flawed RNA and PCR Test Methodology

DID THE CENTRE for Disease Control (CDC) and others correct the problem? Unfortunately, they did not. The CDC continued with the flawed test. Moreover, a test cannot be accepted or used until it is tested against the reference standard of the item (in this case, RNA) it is supposed to test. Therefore, one needs the RNA specimen obtained from the virus. But there is no RNA reference standard available anywhere in the world!

Therefore, there cannot be an RNA test or its associated virus. Any link or conclusion based on the PCR results, cases, epidemiological assessments, lockdown, social restrictions, vaccine developments, and recommendations must be considered scientifically null and void—no exceptions! Otherwise, experts and nations are accepting and promoting one of the greatest scientific frauds.

Following are some more worrying observations with claims to PCR test.

Perhaps, the most noted PCR study protocol in this regard is from Christian Drosten (Charity-University Medicine Berlin, Germany; Public Health England, London, United Kingdom), titled *Detection of 2019 novel coronavirus (2019-nCoV) by real-time RT-PCR*, published in Europe's journal on infectious disease surveillance, epidemiology, prevention and control.[204]

> *According to the World Health Organization (WHO), the WHO China Country Office was informed of cases of pneumonia of unknown aetiology in Wuhan City, Hubei Province, on*

[204] https://www.eurosurveillance.org/content/10.2807/1560-7917.ES.2020.25.3.2000045

> *December 31 2019. A novel coronavirus currently termed 2019-nCoV was officially announced as the causative agent by Chinese authorities on January 7. A viral genome sequence was released for immediate public health support via the community online resource virological.org on January 10 (Wuhan-Hu-1, GenBank accession number MN908947), followed by four other genomes deposited on January 12 in the viral sequence database curated by the Global Initiative on Sharing All Influenza Data (GISAID). The genome sequences suggest presence of a virus closely related to the members of a viral species termed severe acute respiratory syndrome (SARS)-related CoV, a species defined by the agent of the 2002/03 outbreak of SARS in humans. The species also comprises a large number of viruses mostly detected in rhinolophid bats in Asia and Europe.*

The study protocol describes the procedure and the "science" behind the PCR test but, amazingly, in a very confusing way to hide the fundamental flaw of the PCR test and its application. It should be evident that the statement *A viral genome sequence was released for immediate public health support* is incorrect. A virus is not a genome sequence, but RNA assumed to be part of the virus. It is an assumption because the link between RNA and the virus has never been established. The only thing one can say is that sequence is of an RNA (chemical compound) obtained from a small aliquot (25 µL) of the cultured (a treated or thoroughly tortured/disguised) patient swab samples.

Similarly, *The genome sequences suggest the presence of a virus closely related to the members of a viral species termed severe acute respiratory syndrome (SARS)-related CoV* is an incorrect statement regarding the claim of the presence of the virus.

One may ask why such false claims have not been caught and slipped through a peer-review process. Contrary to common belief, the peer-review process is not an unbiased independent third-party review but establishes that claims ("science") made are as taught.

For example, consider the publication timeline where the article was submitted on January 21, 2020, accepted on January 22, 2020 and published on 23 Jan 2020.

Within two days of submitting the article, it got reviewed and published—indicating it was assumed, considering the reputation of the authors ("scientists"), that they must have followed the "accepted science" (incorrect though) that RNA means virus; hence the PCR testing and its conclusion become legit. At this stage, the described PCR test remains equivalent to saying the yellow color of a body fluid containing ultrafine particles indicates that a person might have consumed gold or gold-related food or drink.

The PCR test is not a virus or an illness test. It is an RNA test. Even here, this is an invalid claim. For a test to be valid or accepted for general use, it must be shown working for the item it is supposed to test, in this case, the RNA. Therefore, there has to be a reference RNA sample available that is not. Hence, calling the PCR test an RNA test is also false. Until a reference RNA from the virus is available, a PCR test is not testing RNA, but something else, probably nothing. It is not an opinion but the requirement of fundamental scientific principles and facts.

In scientific terminology, it is said that the test lacks method/test validation, and using a non-validated test reflects a case of (criminal) ignorance and incompetency of the subject by the experts and is often dealt with swiftly by law. A recent example is the shutting down of Theranos lab, which claimed to have developed a blood test.

> *Holmes and Balwani used advertisements and solicitations to encourage and induce doctors and patients to use Theranos's blood testing laboratory services, even though, according to the*

> *government, the defendants knew Theranos was not capable of consistently producing accurate and reliable results for certain blood tests. It is further alleged that the tests performed on Theranos technology were likely to contain inaccurate and unreliable results.*[205]
>
> *Theranos founder Elizabeth Holmes has been convicted of defrauding investors after a months-long landmark trial in California. Prosecutors said Holmes knowingly lied about technology she said could detect diseases with a few drops of blood.*[206]

In scientific terminology, she promoted a non-validated test. The PCR test falls in the same category, i.e., a non-validated swab test to detect a non-existent virus. Regulatory authorities patronize the test based on fraudulent regulatory and CDC sciences. Hence testing is slipping through the cracks, but the end will be the same as of Theranos, noted above. How should a layperson spot a non-validated test or testing? First, there will be reported complaints about the test results being false positives or negatives at random—a common characteristic of the PCR test.

For example, consider the ACCELERATED EMERGENCY USE AUTHORIZATION (EUA.) SUMMARY Modified Thermo Fisher TaqPath COVID-19 SARS-CoV-2 Test (ORF1ab, N, and S gene detection) (Biocerna) EUA Summary Updated: April 6, 2021, downloaded from the FDA site.[207]

---

[205] US v. Elizabeth Holmes, et al., The United States Attorney's Office, Northern District of California. https://www.justice.gov/usao-ndca/us-v-elizabeth-holmes-et-al

[206] https://www.bbc.com/news/world-us-canada-59734254?fbclid=IwAR2Wu8zMJ-FlOKmYQlbEipS-RYXyfwHuNmITtvn1GmyDmyMrb7_ZhnmH-iM

[207] https://www.fda.gov/media/137450/download

The SARS-CoV-2 RT-PCR assay will be performed at Biocerna, certified under the Clinical Laboratory Improvement Amendments of 1988(CLIA), 42 USC §263a as per Laboratory Standard Operating Procedure that was reviewed by the FDA under this EUA.

In the intended use section:

> *The SARS-CoV-2 assay is a real-time reverse transcription polymerase chain reaction (RT-PCR) test intended for the qualitative detection of nucleic acid from SARS-CoV-2 in nasal, mid-turbinate, nasopharyngeal, and oropharyngeal swab and bronchoalveolar lavage (BAL) specimens from individuals suspected of COVID-19 by their healthcare provider. Testing is limited to Biocerna L.L.C., certified under Clinical Laboratory Improvement Amendments of 1988 (CLIA), 42 USC §263a to perform high complexity tests.*
>
> *The SARS-CoV-2 R.N.A. is generally detectable in respiratory specimens during the acute phase of infection. Positive results are indicative of the presence of SARS-CoV-2 R.N.A.; clinical correlation with patient history and other diagnostic information is necessary to determine patient infection status.* ***Positive results do not rule out bacterial infection or co-infection with other viruses. The agent detected may not be the definite cause of disease."*** [Emphasis added] *That is, positive results may not be from the virus or its RNA.*

Further:

> ***Negative results do not preclude SARS-CoV-2***

> ***infection*** *and should not be used as the sole basis for patient management decisions. Negative results must be combined with clinical observations, patient history, and epidemiological information.*

Positive results could be from other reasons (e.g., bacterial infection), not SARS-CoV-2, and negative results do not preclude infection, i.e., it does not mean one has SARS-CoV-2. Hence the PCR test does not mean much—it could come positive or negative at random. In a validated test, positive means SARS-CoV-2 and negative means no SARS-CoV-2. Like if a thermometer shows a higher than normal temperature, the person has a higher temperature, with no ifs and buts.

In conclusion, the PCR test is an invalid and irrelevant test described in the fact sheet for test instructions. What more information would one require to stop recommending or using the test?

The main reason for the unreliability of the PCR test is the non-availability of the pure isolated virus or it's RNA (reference/gold standard).

The medical and clinical experts (CDC science) promote this as an accurate and reliable (gold standard) test: Scientists use the PCR technology to amplify small amounts of RNA from specimens into deoxyribonucleic acid (DNA), which is replicated until SARS-CoV-2 is detectable if present. The PCR test has been the gold standard test for diagnosing COVID-19 since authorized for use in February 2020. It's accurate and reliable.[208]

But, with the new announcement from the CDC requiring withdrawal of the PCR test:

> *After December 31, 2021, CDC will withdraw*

[208] https://my.clevelandclinic.org/health/diagnostics/21462-covid-19-and-pcr-testing

> *the request to the U.S. Food and Drug Administration (FDA) for Emergency Use Authorization (EUA) of the CDC 2019-Novel Coronavirus (2019-nCoV) Real-Time RT-PCR Diagnostic Panel, the assay first introduced in February 2020 for detection of SARS-CoV-2 only. CDC is providing this advance notice for clinical laboratories to have adequate time to select and implement one of the many FDA-authorized alternatives.*[209]

So the gold standard PCR test, approved or authorized by CDC, would not be available or used after December 31, 2021, but is it?

> *In preparation for this change, CDC recommends clinical laboratories and testing sites that have been using the CDC 2019-nCoV RT-PCR assay select and begin their transition to another FDA-authorized COVID-19 test. CDC encourages laboratories to consider adopting a multiplexed method that can facilitate the detection and differentiation of SARS-CoV-2 and influenza viruses. Such assays can facilitate continued testing for both influenza and SARS-CoV-2 and can save both time and resources as we head into influenza season. Laboratories and testing sites should validate and verify their selected assay within their facility before beginning clinical testing.*[210]

---

[209] https://www.cdc.gov/csels/dls/locs/2021/07-21-2021-lab-alert-Changes_CDC_RT-PCR_SARS-CoV-2_Testing_1.html

[210] Ibid.

Chapter 30—United Kingdom COVID PCR and LFD Tests

THE UNITED KINGDOM said from Friday, Jan 7th, 2022, it will no longer require vaccinated travelers to take COVID tests before boarding flights to the country.[211]

According to Boris Johnson, UK prime minister at the time, the requirement for travelers to self-isolate on arrival until receiving a negative polymerase chain reaction (PCR) test would also be scrapped from Sunday, Jan 9th, 2022. Instead, new arrivals will be required to take a lateral flow test no later than day two.

If a traveler tests positive with a lateral flow, they will then be required to take a PCR test and self-quarantine.

However, the rule appears to have been changed to:

> *Under the new guidance, anyone in England who receives a positive lateral flow device (LFD) test result should report their result on gov.uk and must self-isolate immediately, but will not need to take a follow-up PCR test.*[212]

The PCR test as an invalid test, as noted in the previous chapter.

Notably, the PCR test is still (July 6, 2022) actively prescribed with the same claim:

> *Shown to be highly accurate, the PCR lab test is the gold-standard of COVID-19 testing.*[213]

Rather than moving away from the PCR, as noted above, for an

[211] https://www.thecable.ng/explainer-what-is-covid-lateral-flow-uks-replacement-for-pcr-arrival-test

[212] https://www.bbc.com/news/uk-59878823

[213] https://www.ontariohealth.ca/COVID-19/testing-analysis

unknown reason, the FDA introduced a different test (Home/Rapid or Antigen test).

It is important to note that Home/Rapid/Antigen/Lateral-Flow Device tests are one and the same thing. However, they are marketed under different names.

> *To put it simply—there is no difference between a rapid antigen and a lateral flow test! Rapid testing uses the well-established lateral flow technology to detect proteins or antigens that are present when a person is infected with COVID-19. The FDA considers at-home COVID-19 diagnostic tests to be a high priority, and we have continued to prioritize their review given their public health importance.*
>
> *Most antigen tests for at-home use are authorized for serial testing, or testing the same individual more than once within a few days. These authorizations followed the announcement of a streamlined approach to help facilitate the authorization of rapid tests for use with serial testing programs, which has increased consumer access to testing.*
>
> *The FDA wants to remind patients that all tests can experience false negative and false positive results. Individuals with positive results should self-isolate and seek additional care from their health care provider. Individuals who test negative and experience COVID-like symptoms should follow up with their health care provider as negative results do not rule out a COVID-19 infection.*[214]

---

[214] https://www.nomadtravel.co.uk/blog/travel-health/rapid-antigen-test-vs-lateral-flow

What is a lateral flow test? A lateral flow test is a quick/fast test for COVID-19 that provides results in 30 minutes. It is recommended for people who do not have symptoms of the virus. The test kit is a hand-held device with an absorbent pad at one end and a reading window at the other.

A sample of mucus is taken from a person's nose or throat using a swab. The swab is then dipped in a tube containing a solution to dilute the sample and then dripped onto the device's paper pad. A colored band will appear on the control zone to show the test has worked. A colored band also appears on the test zone labeled, indicating a positive test.

How is it different from a PCR test? PCR tests detect the virus' genetic material, i.e., RNA, while lateral flow tests work by detecting protein (spike protein) from the virus.

Also, although both tests require one to take a swab from the nose or throat, PCR tests must be sent to a laboratory, while lateral flow tests can easily be done at home and give results in minutes.

Why is it recommended for people without symptoms? According to the NHS, about one in three people with COVID do not have symptoms but can still infect others. The rapid lateral flow tests help to find positive cases earlier and break hidden transmission chains.

How effective is the lateral flow test for covid? Research shows that rapid tests are reliable for detecting the coronavirus. A recent study showed that when used correctly, lateral flow tests are likely to have sensitivity above 80% and, in many cases, above 90%. How was the sensitivity

> *and accuracy of the lateral flow test established? Such studies are usually done by comparing the results against the PCR test, the same PCR testing which is considered problematic, stating (r = 0.968).*[215]

Scientifically, one does not compare method vs. method but against a reference. If the method is considered a standard, it must first be validated against the reference, which is RNA or the virus.

However, as the RNA or virus is not available, one cannot validate a method itself or compare it to another method. This clearly shows a lack of understanding of tests or method development and validation expertise, in general science, in the medical area.

Importantly, this Rapid, Antigen or LFD test is known to be flawed and weaker than the PCR from the beginning of the pandemic. For example:

> *On Friday (Nov 13, 2020), Elon Musk said that in one day, he took four Covid-19 tests—called rapid antigen tests—and received two negative and two positive results. He also tweeted that he took two PCR tests at separate labs and is waiting for the results.*

Musk, who has been dismissive about COVID-19 and its severity, tweeted that something "extremely bogus" was going on.

Furthermore, in the same article, it is clearly described that:

> *Although these types of tests are cheap to manufacture (they can cost $25-$100 out of pocket) and can deliver test results in a matter of*

215 https://hospitalhealthcare.com/covid-19/covid-19-lateral-flow-tests-comparable-to-pcr/

> *minutes, they are not as reliable as a polymerase chain reaction, or PCR...*[216]

Still, FDA and CDC are heavily promoted this scientifically unreliable test.

Understandably, it is a very confusing situation, particularly for a common person. It appears that it is intentional to fool people into fear and coercion to accept vaccination and its multiple boosters.

Translating the whole jargon into simple language would be that the PCR test did not work. So, another test is suggested, which determines a protein from the virus instead of the RNA. However, there is no reference standard available for the protein. Therefore, scientifically speaking, this method will be invalid as well. Therefore, it cannot be a relevant and reliable test either. However, at the same time, PCR tests continue, which do not have scientific merit.

Confusion persists, not the science.

---

[216] https://www.cnbc.com/2020/11/13/musk-tests-positive-and-negative-for-covid-19-antigen-vs-pcr-tests.html

Chapter 31—North American COVID-19 Reactions

CONSIDER THE RISE in the recent number of "cases."

> *While Canada could see a sharp peak and decline in cases in the coming weeks, given disease activity far exceeding previous peaks, even the downside of this curve will be considerable," said Dr. Theresa Tam, Canada's chief public health officer.*[217]
>
> *The new year is starting with a massive influx of Covid-19 that's different from any other during this pandemic, doctors say: "We're seeing a surge in patients again, unprecedented in this pandemic," said Dr. James Phillips, chief of disaster medicine at George Washington University Hospital.*[218]
>
> *The United States leads the world in the daily average number of new infections reported, accounting for one in every 4 infections reported worldwide each day. The United States is reporting 738,047 new infections on average each day, 92% of the peak—the highest daily average reported on January 14.*[219]

---

[217] https://globalnews.ca/news/8511154/omicron-canada-covid-modelling/

[218] https://www.cnn.com/2022/01/02/health/us-coronavirus-sunday/index.html

[219] https://graphics.reuters.com/world-coronavirus-tracker-and-maps/countries-and-territories/united-states/

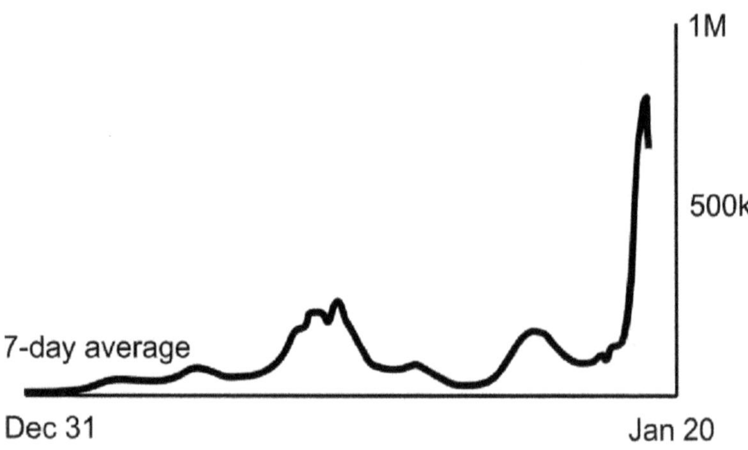

Reproduced from:https://graphics.reuters.com/world-coronavirus-tracker-and-maps/countries-and-territories/united-states/

In addition, The US government is providing an extraordinary number of Rapid Tests, which, just on a probability basis would provide an increase in test positivity.

For example, suppose 1 billion tests are to be done, and considering 1 in 5 would give false positives as one would expect. In that case, there will be 200 million positive tests (cases) based on the expected variability of the sheer number of tests. The lack of validation of the test would add more numbers to it. This creates false higher numbers of cases.

The remarkably similar trend has been reposted for Canada.[220]

---

[220] https: / / www.biopharminternational.com / view / fda-recommends-pausing-use-of-j-j-covid-19-vaccine-after-reports-of-blood-clots

Chapter 32—Omicron Variant and Rapid Antigen/Home Tests

THE OMICRON VARIANT has been blamed for the surge, a non-existent variant of a non-existent virus, as none are isolated and characterized. It strongly suggests that an increase in positive test results (cases) may be a reflection of change to the alternative (rapid antigen) test.

It is often mentioned that the virus has been isolated and is commercially available from ATCC [221] and BEI Resources.[222]

It is not an accurate statement. These companies do not sell the virus or its variants but isolate or lysate with deceptive labeling. For example, looking at the description of one of the virus's variants, i.e., heat-inactivated SARS-CoV-2, Delta variant, under detailed product information/Comments, it says: *This isolate is lineage AY.24* and *the following mutations are present in the clinical isolate.* [223]

Note the word isolate, which is cell culture/gunk, not the isolated virus. Under shipping, the information says:

> *Each vial contains approximately 0.25 mL of heat-inactivated, clarified cell lysate and supernatant*

Note the word lysate, which is the soup from the breakup of cells in medium/culture, not the virus.

For $1200, what's the customer really buying? A diluted human mucus/phlegm from swab samples with all kinds of added chemicals (30+), including African green monkey kidney cell (Vero cells) broth, not the isolated or pure virus, as the trusting people/customers believe.

---

[221] https://www.atcc.org/

[222] https://www.beiresources.org/Home.aspx

[223] https://www.atcc.org/products/vr-3342hk

Similarly, looking at the BEI Resources site, there are about 40+ SARS-CoV-2 virus-related entries.[224] All are labeled as "SARS-CoV-2 isolate," not a single one with SARS-CoV-2 virus because no isolated virus is available.

BEI Number	Description	Clinical Information Available
NR-56461	SARS-CoV-2, **Isolate** hCoV-19/USA/MD-HP20874/2021 (Omicron Variant)	Isolated from a nasal swab from a human as part of surveillance testing on November 27, 2021, in Maryland, USA
NR-55607	SARS-CoV-2, **Isolate** hCoV-19/Uganda/MUWRP-20200195568/2020 (Delta Variant)	Isolated from a human nasal swab in Uganda on December 28, 2020
NR-55691	SARS-CoV-2, **Isolate** hCoV-19/USA/CA-VRLC086/2021 (Delta Variant)	Isolated from a human mid-turbinate nasal swab in California, USA, on June 21, 2021
NR-55672	SARS-CoV-2, **Isolate** hCoV-19/USA/MD-HP05647/2021 (Delta Variant)	Isolated in Maryland, the USA, on April 27, 2021

Table 2

First four listings of available SARS-CoV-2 "ISOLATES" from BEI

In the product details section, it is noted that it is an "isolate" of Severe Acute Respiratory Syndrome-related coronavirus 2 (SARS-CoV-2). Isolate hCoV-19/USA/MD-HP20874/2021 was isolated from a nasal swab in Maryland, USA, on November 27, 2021.[225]

In short, they are faking it and lying with confidence and authority. It is a clear case of fraudulent business practice.

No wonder Costco, Walmart, Amazon, etc., retailers of the world, do not sell this stuff. They will be behind bars immediately for making such false claims and marketing. Remember the

---

224 https://www.beiresources.org/BEIHighlights1.aspx?ItemId=79&ModuleId=14004

225 https://www.beiresources.org/Catalog/animalviruses/NR-56461.aspx

Theranos case described earlier.[226]

Only pharmaceutical manufacturers can make such false and fraudulent claims as they are usually protected by the regulatory authorities' patronage and approval. No third independent audit is available to check and control this deceitful activity.

In short, no one has isolated, purified, and characterized the virus. Therefore, it is not available from anywhere.

This false description or labeling is not unintentional or an oversight. Instead, such claims or misrepresentations are based on cleverly worded reports published in journals. For example, reading from a "scientific" article:

> *Our results contribute to the understanding...the first detailed view of coronavirus ultrastructure.*

It's a similar material and method approach used in the CDC study described above. However, for convenience and clarity, below is the text reproduced from the Material and Methods section.

> *Virus growth and purification. Vero-E6, feline AK-D, and murine DBT cells were maintained in Dulbecco's modified Eagle medium (Invitrogen) supplemented with 10% fetal bovine serum (HyClone), 2 mM L-glutamine, and antibiotics. SARS-CoV-Tor2 was propagated on Vero-E6, FCoV-Black was propagated on AK-D, and MHV-OBLV60 was propagated on DBT cells. Viral supernatants were collected 24 h postinoculation, and infectious titers were determined by plaque assay as described previously. Virions were harvested from infected cell culture medium by*

[226] https://www.bbc.com/news/world-us-canada-59734254?fbclid=IwAR16xCEplp2invQgE37PKIKiyssPDHPOHJjK3ZTpAnvG2oKSz9kyxF2WMxo

> *precipitation with 10% polyethylene glycol (PEG-8000). The PEG pellet was resuspended in 0.9% NaCl with 10 mM HEPES and applied to a 10 to 30% sucrose density gradient. Final purification was achieved by ultracentrifugation at 27,500 rpm for 2 h at 5°C. Virus particles resuspended in HEPES-buffered saline were fixed with 10% (SARS-CoV) or 1% (FCoV and MHV) formalin overnight at 4°C prior to cryopreservation.*[227]

The use of the wording *Final purification was achieved...*compels readers to believe that purification is being done and the virus has been isolated, which is an incorrect description of the methodology described. In reality, the text describes mixing (adding, not separating, or isolating) the swab sample with numerous chemical reagents, including African green monkey kidney cells. Further, virus particles were resuspended (mixing and adding) in more chemical reagents (HEPES-buffered saline) following ultracentrifugation.

The question is: how did it get established that particles represent the virus and SARS-CoV-2? This is what is supposed to be determined. But unfortunately, one cannot even see the particles with the naked eyes; hence, electron microscopy. And the sample is still in solution or suspension form, not the pure and single component.

In reality, swab samples were treated with many chemical reagents and cell components, (ultra) centrifuged, and aliquoted. It is assumed that liquid contains particles considered SARS-CoV-2 without evidence mixed (resuspension) with more chemicals. Pictures of solution or suspension (the "isolate") were taken, and some random circles were highlighted or marked as the virus.

Nothing has ever been isolated or purified.

Medical professionals and experts should stop using such

---

[227] https://journals.asm.org/doi/epub/10.1128/JVI.00645-06

deceptive language to mislead the public and science that the virus has been isolated and purified. It is the distortion of facts and science at its highest level.

Chapter 33—Former NIAID Director Dr Anthony Fauci; Vaccine Opinions

VACCINES HAVE BEEN developed and mass-marketed to protect the public from the virus and its ill effects, often called illness, infection, hospitalization (including ICU admissions), and deaths. Vaccines are claimed to be highly effective, showing an efficacy of 95%.

> *Clinical US trials of the Pfizer/BioNTech and Moderna vaccines showed 95% and 94.1% efficacy,* [Anthony Fauci] *said.*
>
> *Anthony Fauci, MD, told attendees at the ATS 2021 International Conference that the real-world evidence of vaccines against SARS-CoV-2, the virus that causes COVID-19, is even better than expected.*
>
> *The head of the National Institute of Allergy and Infectious Diseases and chief medical advisor to President Joe Biden also reiterated the administration's goal of having 70% of US adults inoculated with at least 1 dose of a COVID-19 vaccine by July 4.*[228]

Understandably, people will believe that what experts describe is accurate and based on the science claimed by "scientists."

The claim is that the efficacy is 95%. The question (or science part) is how one calculated the efficacy. From the CDC website,

[228] https://www.ajmc.com/view/efficacy-of-covid-19-vaccines-in-real-world-settings-even-better-than-expected-fauci-says

vaccine efficacy (VE) is interpreted as the proportionate reduction in disease among the vaccinated group. So a VE of 95% indicates a 95% reduction in disease occurrence among the vaccinated group or a 95% reduction from the number of cases you would expect if they have not been vaccinated.[229]

However, it is RVE (Relative Vaccine Efficacy), not the VE (Vaccine Efficacy) as described or the public assumes. The VE means one should treat an equal number of patients in both groups (non-vaccinated and vaccinated) and see how many reacted/cured in the vaccinated group. However, vaccine efficacies have never been determined in patients (subjects having virus or infection). Therefore, getting actual and accurate efficacy data from vaccination is impossible.

Secondly, the claim that *the real-world evidence of vaccines against SARS-CoV-2, the virus that causes COVID-19...*is scientifically invalid because the vaccines have never been tested against SARS-CoV-2. There is no sample/specimen of SARS-CoV-2 available or used, as explained in detail in the science section. Furthermore, studies (clinical trials) have never used the virus or patients with the virus. Therefore, claims of established efficacy are invalid on this basis as well.

Thirdly, vaccine efficacy has been tested in healthy human subjects against the PCR test, i.e., if healthy subjects in the vaccinated group show relatively lower positive results than the non-vaccinated group—it is considered the vaccine worked or efficacious. However, as explained above, the PCR test has no relationship with the virus, infection, or illness (i.e., COVID-19).

Therefore, the irrefutable conclusion is that the efficacy of the vaccines has never been established, let alone the high efficacies of 94%+ as claimed. Therefore, these are scientifically invalid and false claims—i.e., nowhere science be found. These are just claims for the sake of claims.

In short, healthy subjects were given the vaccines by injection.

---

[229] https://www.cdc.gov/csels/dsepd/ss1978/lesson3/section6.html

They were monitored with the PCR test to indicate the vaccine's effectiveness, i.e., if tested negative (vaccines worked or efficacious, otherwise not). One wonders what kind of science this is. The test never determines or is validated against any virus, illness, or disease but would tell the presence or absence of these things. There is no other word for such a practice but con artistry and fraud—certainly, clinical trials and the claims were not scientific by any measure.

Some thoughts about messenger (mRNA) vaccines, which are so heavily promoted as modern-day wonder therapeutics or technology, as being thoroughly studied for a decade claimed by scientists and experts.

For example, the CDC site says:

> *Researchers have been studying and working with mRNA vaccines for decades. Interest has grown in these vaccines because they can be developed in a laboratory using readily available materials. This means vaccines can be developed and produced in large quantities faster than with other methods for making vaccines.*[230]

Dr. Stephen Hoge, Moderna, Inc., President, expressed a similar view stating:

> *Since 2010, we have built and invested in our technology platform, which creates mRNA sequences that cells recognize as if they were produced in the body.*

However, translating it into a simple language, means synthesizing chemical compounds (sequences). Hence, this would be the first

[230] https://www.cdc.gov/coronavirus/2019-ncov/vaccines/different-vaccines/mrna.html

product when developed and promoted as new and modern technology.

This will be a synthetic and unnatural (foreign to the body compound) that has never been tested in humans before, and the body has to deal with it, as explained in the science section of the chapter. Therefore, one needs to be extra cautious about its safety profile.

However:

> *White House coronavirus advisor Dr. Anthony Fauci said Monday he is "not particularly concerned" about the safety risk of a potential coronavirus vaccine by Moderna, despite the fact that it uses new technology to fight the virus.*

Strange, no safety assessments or worries. This is against all commonly accepted and required standards, regulatory practices, and requirements.

Indeed, no safety studies were conducted, as noted by Moderna's President in his testimony to Hearing before the House Energy and Commerce Committee Subcommittee on Oversight & Investigations Testimony of Dr. Stephen Hoge President, Moderna, Inc. July 21, 2020:

> *In our prior work on betacoronavirus mRNA vaccines, we identified a key protein on the surface of coronaviruses, called the Spike protein, as a good vaccine candidate.*
>
> *As I noted above, we began work on mRNA-1273 immediately after the genetic sequence of the novel coronavirus was released on January 11, 2020. Only 25 days later, on February 7, 2020, Moderna completed its first clinical batch of mRNA-1273.*
>
> *The Phase 1 study, led by NIH, dosed its first*

> *participant on March 16, 2020. On May 18, 2020, we announced positive interim results from the mRNA-1273 Phase 1 study, which showed the generation of neutralizing antibody titer levels in all eight initial participants.* [231]

There is no mention of animal or in vitro safety or toxicity assessment studies. It is not clear how scientists and regulatory experts allowed such negligence. And within two months of producing the mRNA, the phase 1 clinical study was complete. If NIH (government authority) had not led the study, such a study would certainly fall in the category of scientific and medical fraud. But the scientists proceeded with it anyway.

> *Unlike traditional vaccines, which expose the body to a viral protein to stimulate the immune system, mRNA acts as an instruction kit, telling the body how to construct the proteins itself. The immune system then responds to the viral protein by making antibodies.*[232]

How and where this spike protein was isolated from the virus (SARS-CoV-2)? It has never been explained. No isolated and purified protein reference standard is available from any source, and it cannot be because the virus has never been isolated. There is no evidence, not a single one, in the literature that the virus has been isolated and then the (spiked) protein from it.

If one summarises the CDC science version of the COVID-19

---

231 https://energycommerce.house.gov/sites/democrats.energycommerce.house.gov/files/documents/Testimony%20-%20Hoge%2020200721_0.pdf

232 https://khn.org/news/fauci-unfazed-as-scientists-rely-on-unproven-methods-to-create-covid-vaccines/

pandemic and virus narrative, the following will be an example, First, *assume* a pandemic and then *assume* that people are getting sick and dying (starting point). Then "confirm" the pandemic spread based on random evaluation of hospital visits, mostly one patient per hospital/country. If most patients evaluated were successfully treated with common and standard treatments under the situation and recovered without serious illness or deaths, still *assume* that countries are in a deadly pandemic.

Next, *assume* that the infection is virus-based and *the virus is novel, call it SARS-CoV-2,* "confirmed" on a PCR test. The test has not been shown to work for the virus or its illness (because it cannot test them), but it still *assumes* that the test works for the novel virus. It is unnecessary to validate the test against the reference standard but *assume* it is validated. To provide "scientific" proof of the novel virus's existence, conduct isolation of the "isolate" and *assume* that isolate/lysate is the virus. Once labeled with the virus infection (i.e., PCR positive), isolate the patients from others and *assume* they will recover. If not recovered, then *assume* that they died of virus illness or COVID-19.

Further, *assume* that the only effective treatment has to be a vaccine and *assume* it has to be a new one. *Assume* no current medicine or therapy is workable. Conduct clinical trials in healthy volunteers (not patients sick with the virus). However, *assume* that with the PCR-test negative results, subjects got protected from the virus, which was *assumed* to be present. Calculate the RVE (Relative Vaccine Efficacy), not real or absolute vaccine efficacy, to *assume* that the vaccines' efficacy is real and they have been highly successful.

One wonders, how has science been followed? There is no science but assumptions. Who approved this science—the peers and peers dominated regulatory authorities and expert advisory committees?

Moving along, it was well established that facemasks are not useful in protecting from the virus.

At the beginning stage of the pandemic, Dr. Fauci said:

> *There's no reason to be walking around with a mask. When you're in the middle of an outbreak, wearing a mask might make people feel a little bit better, and it might even block a droplet, but it's not providing the perfect protection that people think that it is. And, often, there are unintended consequences—people keep fiddling with the mask and they keep touching their face.*

A tweet on Feb 27, 2020, from CDC, said:

> *...the CDC does not currently recommend the use of facemasks to help prevent novel #coronavirus. Take everyday preventive actions, like staying home when you are sick and washing hands with soap and water, to help slow the spread of respiratory illness.*[233]

This position was changed a week later, on April 3, 2020:

> [CDC] *updated its previous advice and recommended people wear cloth face coverings "in public settings when around people outside their household, especially when social distancing measures are difficult to maintain."*

However, denying or rationalizing this policy flip-flop, views were presented "As more information became available about SARS-CoV-2, the virus that causes COVID-19, health authorities and organizations around the world have changed their stance towards the impact of face masks and the spread of the disease.

Fact check: outdated video of Fauci saying "there's no reason

[233] https://mobile.twitter.com/cdcgov/status/1233134710638825473

to be walking around with a mask." [234]

And now a more recent view from a medical expert (Dr. Leana Wen):

> *CNN medical analyst supports Democrat-led states dropping school mask mandates: 'The science has changed.'* [235]

Wow, science changed—again! Sorry, science did not change—lies changed.

Clearly, there is a change in the recommendation of face mask use. There is no harm in changing recommendations or policies if and when newer data becomes available. However, no supporting scientific data is provided for masks on or off recommendations—when technically and scientifically, it is easy to conduct a scientific experiment to prove one way or another.

Scientifically speaking, the usefulness of wearing face masks can be established with a straightforward laboratory experiment.

For example, one may use a two-chamber container separated by a mask holder (shown in the picture below), providing airflow with virus from one side to the other. By sampling both sides of the chamber and measuring the amounts of virus, one can quickly determine the virus-holding capability of the mask or any other filtering media.

However, no such experiment has been reported or planned to be conducted.

---

[234] https://www.reuters.com/article/uk-factcheck-fauci-outdated-video-masks-idUSKBN26T2TR

[235] https://www.foxnews.com/media/cnn-medical-analyst-changes-tune-covid-restrictions-science-changed

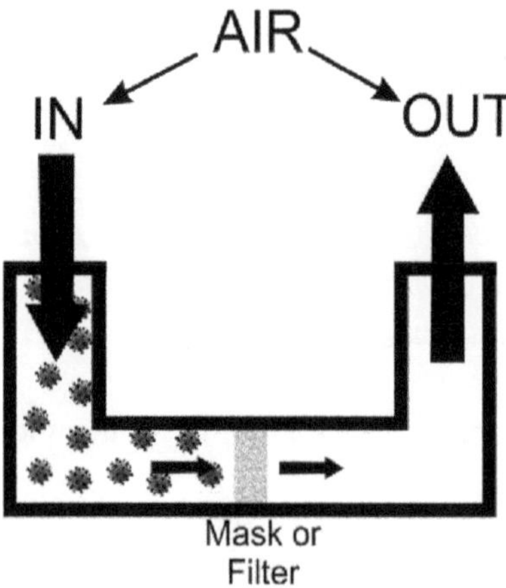

Figure 39
Schematic of a Possible Tester for Evaluating Face Mask Effectiveness

The reason is such an experiment requires some quantity of actual virus and a test method to quantify the virus. There would not be any human subjects requirement, i.e., no human toxicity or safety concerns. This would be a simple high school or undergraduate level laboratory testing which could be done within days. However, this is where the problem is!

Experts do not tell the public that both of these, i.e., actual virus and a valid test to monitor the virus, are unavailable, which will collapse the whole virus and pandemic issue. The public will know or question that if there is no actual virus and its valid test available, then what is the basis of claiming the virus's existence and its associated disease (COVID-19)? There will not be any answers.

Similar simple experiments can also be devised to establish the usefulness of sanitizers, handwashing, and social distancing. However, none of these experiments can be conducted either because they require a valid test and a reference virus sample.

The medical/CDC science pretends to provide supporting claims for mask use based on studies such as one "Mayo Clinic

Minute: Study shows masks can prevent COVID-19." [236]

The study title is *Combined Effects of Masking and Distance on Aerosol Exposure Potential* and it states:

> *To quantify the efficacy of masking and "social distancing" on the transmission of airborne particles from a phantom aerosol source (simulating an infected individual) to a nearby target (simulating a healthy bystander) in a well-controlled setting.*[237]

It assumes aerosol as a virus or has the virus and mannequin as humans on the street and conclusion about preventing COVID-19. It reflects an astonishing lack of understanding of the subject of science, conducting its experimentation and drawing conclusions from them from a considered highly reputed medical organization, Mayo Clinic.

The medical community should note that such studies are considered scientific or experimental. However, they are not. These are folklore-type observational or guessing. No virus specimen is used in the experiments, and no scientifically valid test method is available to test the virus. So how could such studies provide an answer to the mask's effectiveness? They cannot.

It would exactly like the development and testing of hospital beds where clinical trials may be used to assess their performance and durability by monitoring patients' health/illness under the supervision of physicians. Certainly, physicians and patient can provide their input as the users for comfort or discomfort of the beds, but development and testing will remain the responsibility of

---

[236] https://newsnetwork.mayoclinic.org/discussion/mayo-clinic-minute-study-shows-masks-can-prevent-covid-19/

[237] https://www.sciencedirect.com/science/article/abs/pii/S0025619621004018?dgcid=author

the beds' manufacturer. They have been trained for such, not the physicians.

From the scientific perspective, it is impossible to develop a treatment for a disease with no measurable parameter and validated method to monitor it. COVID-19 falls in this category. It is claimed that the virus (SARS-CoV-2) caused the disease; however, no validated method is available to monitor the virus. Furthermore, the PCR test does not monitor the virus but a chemical compound called RNA. This is a blatant example of ignorance of the medical profession's chemistry/science. Based on chemical science understanding, it was clearly described that vaccines or therapeutic are not possible.

> *Arguably, there appears to be no need, at least on an urgent basis, for developing a vaccine or any other new therapies for the illness showing mild flu-like symptoms, which could be handled with already developed and available medications. Clinical trials have been conducted without scientifically valid study designs based on vague endpoints and invalid analytical (PCR) tests that ought to produce useless conclusions and products.*[238]

In continuation on this aspect, it was described by the author in another article, published on August 6, 2020, before developing vaccines:

> *On the other hand, it is impossible to develop a true vaccine for COVID-19 because—if we are*

[238] https://principia-scientific.com/newsletter-latest-from-dr-judy-wilyman-phd/ Link to the original blog article: https://bioanalyticx.com/should-fda-and-other-authorities-approve-the-sars-cov-2-covid-19-vaccines-a-scientific-perspective/

incapable of actually monitoring the virus or disease—how can the effectiveness of a vaccine be established?

In short, it cannot!

Therefore, most likely, a fake vaccine will be developed to satisfy the regulators' wishes—as well as to calm down the created public hysteria and fear. Unfortunately, such vaccines, if developed and administered, will certainly create potentially dangerous side effects, without any presumed benefits, by interfering with the body's own immune system, as well as other related physiological processes.[239]

Clearly, this prediction turned out to be 100% correct. Unfortunately, the vaccines used do not provide the suggested protection or cure. The case numbers still increase but are now called breakthrough cases. What does a breakthrough case means—protection (fence) the vaccines were supposed to provide could not provide, and the virus broke through the fence. The breakthrough cases are, in fact, a hideous description of the failure of vaccines to fool the public.

The CDC website describes under *The Possibility of COVID-19 after Vaccination: Breakthrough Infections*:

…since vaccines are not 100% effective at preventing infection, some people who are fully vaccinated will still get COVID-19.[240]

---

[239] https://principia-scientific.com/covid-19-vaccine-not-possible-for-a-virus-not-yet-identified/

[240] https://www.cdc.gov/coronavirus/2019-ncov/vaccines/effectiveness/why-measure-effectiveness/breakthrough-cases.html

On July 27, 2021, CDC released updated guidance on the need for urgently increasing COVID-19 vaccination coverage and a recommendation for everyone in areas of substantial or high transmission to wear a mask in public indoor places, even if they are fully vaccinated. CDC issued this new guidance due to several concerning developments and newly emerging data signals.

Arguably, rather than considering or admitting the lack of effectiveness of the vaccines, an abrupt twist in the narrative was made, i.e., vaccines are effective, but the virus has mutated to a different version (called "Delta"), stating:

> *...new data began to emerge that the Delta variant was more infectious and was leading to increased transmissibility when compared with other variants, even in some vaccinated individuals.*

This is described under the heading of *Delta Variant: What We Know About the Science.*[241] Note the use of the word "science." It is not science, but CDC science—the arbitrary, self-proclaimed, and fake—CDC-science one.

> *The Delta variant causes more infections and spreads faster than early forms of SARS-CoV-2, the virus that causes COVID-19.*
>
> *Fully vaccinated people with Delta variant breakthrough infections can spread the virus to others. However, vaccinated people* ***appear*** *to spread the virus for a shorter time: Vaccines in the US are highly effective, including against the Delta variant.* [emphasis added]

This is interesting.

---

[241] https://www.cdc.gov/coronavirus/2019-ncov/variants/delta-variant.html

The original version of the virus mutated into a new one (variant). How is it established as a new variant, not a completely different virus? There is no test available for determining the SARS-CoV-2 virus, let alone the one that would differentiate between a new and old one. Further, how is it established that the vaccines are working (highly effective) against the new (Delta) variant? There is no information on a clinical study supporting claims of the Delta variant and the effectiveness of vaccines against it. Simple logic dictates the vaccines were ineffective as predicted and needed to be stopped. The public and patients did not expect to be given excuses. Vaccination continues along with the increase in breakthrough cases. So the next excuse was introduced—a newer variant called Omicron.

Concerning the appearance of Omicron in the United States:

> *CDC is working with state and local public health officials to monitor the spread of Omicron. As of December 20, 2021, Omicron has been detected in most states and territories and is rapidly increasing the proportion of COVID-19 cases it is causing.*[242]
>
> *CDC has been collaborating with global public health and industry partners to learn about Omicron as we continue to monitor its course. We don't yet know how easily it spreads, the severity of illness it causes, or how well available vaccines and medications work against it.*

On the same page, it is described:

> *Current vaccines are expected to protect against*

[242] https://www.cdc.gov/coronavirus/2019-ncov/variants/omicron-variant.html

> *severe illness, hospitalizations, and deaths due to infection with the Omicron variant.*

They (CDC) are learning about the Omicron, however, makes claim that vaccines are expected to be effective against the Omicron." Science would never make such disjoint claims.

Perhaps the more disturbing is the claim that:

> *With other variants, like Delta, vaccines have remained effective at preventing severe illness, hospitalizations, and death. The recent emergence of Omicron further emphasizes the importance of vaccination and boosters.*

But, vaccines were never developed or tested against severe illness, hospitalizations, and deaths. These were not the parameters monitored during the vaccine development or established efficacy. Instead, vaccines' effectiveness was based on the PCR test outcomes, indicating a false claim in support of vaccination efficacy. This is certainly, an invalid scientific claim.

Concerning the vaccine's efficacy, in a recent statement from CDC, it is described:

> *In its latest (27 Dec 2021) official guidelines, the CDC announced that "the efficacy against infection for a two-dose mRNA vaccine is approximately 35%!"* [243]

This is another example of the assumptions explained earlier in "medical science"/virology. The reduction in numbers (efficacy) appears in hospital data. The actual number of volunteers or patients involved is not published and known. One requires (two

---

[243] https://www.cdc.gov/media/releases/2021/s1227-isolation-quarantine-guidance.html

PCR-positive) numbers (1) positive with vaccination (2) positive without vaccination. Let us assume one gets 130 positives with vaccination and 200 for unvaccinated for this discussion. The RVE will be 35% [{(200-130)/200}*100], hence a 60% reduction from the earlier 95%. They are finding relatively more positives with vaccination, thus, presumed reduction in efficacy.

Pfizer CEO (Mr. Bourla) expressed a similar view stating:

> *...the two-dose vaccine does not provide robust protection against infection, and its ability to prevent hospitalization has also declined.*[244]

In the real world, such data would be considered a failure of the vaccination, which indeed it is. Instead, it is promoted as a need for boosters in virology or vaccination. Bizarre.

Authorities are making claims along the way without any valid scientific experimentation, clearly demonstrating a critical lack of understanding and training in science.

Almost three years into the "pandemic" and two years after the vaccination, the number of cases has increased irrespective of the vaccination status.

The chart below from WHO summarizes the number of COVID-19 cases, death, and vaccine doses for the USA.[245]

---

[244] https://www.cnbc.com/2022/01/10/pfizer-ceo-says-two-covid-vaccine-doses-arent-enough-for-omicron.html

[245] https://covid19.who.int/region/amro/country/us

In United States of America, from 3 January 2020 to 4:52pm CET, 19 January 2022, there have been 66,254,888 confirmed cases of COVID-19 with 846,647 deaths, reported to WHO. As of 14 January 2022, a total of 511,548,471 vaccine doses have been administered.

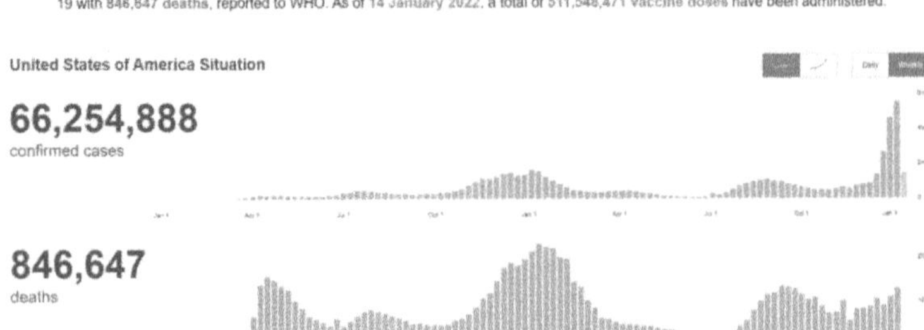

Figure 40
Counts of the Number of Cases and Deaths Allegedly Linked to COVID-19

The number of COVID-19 deaths by age (US), as per the CDC site, clearly shows that the death count has increased during the year 2021 (the vaccination year) compared to the previous year (2020) without vaccination.[246]

This increased number of deaths indicates a lack of effectiveness of the vaccination approach, instead arguably causing harm.

Age Group	Year (2020)	Year (2021)
45 and under	10167 (2.6%)	25147 (5.6%)
45 - 64	63900 (16.6%)	115138 (25.5%)
65 and above	311377 (80.8%)	311162 (68.9)
Total	385444	451447

However, concerning the adverse effects of COVID-19 vaccines, an alarming trend emerged as reported by the CDC database VAERS (The Vaccine Adverse Event Reporting Results).

[246] https://www.cdc.gov/nchs/covid19/mortality-overview.htm

Event Category	Events reported	Percent
Deaths	30,479	2.19
Life Threatening	33.705	2.42
Permanent Disability	56,994	4.40
Congenital Anomaly / Birth Defect	1143	0.08
Hospitalized	174,371	12.54
Existing Hospitalization prolonged	1997	0.14
Emergency Room /Office Visit	121	0.01
Emergency Room	134,126	9.65
Office Visit	230.792	14.66
None of the above	907,415	65.25
Total	**1,544,143**	**11104**

Table 3
The VAERS Results[247]

Table 3 provides evidence that vaccination is causing serious harm to people. Yet, surprisingly, regulatory authorities are not alarmed by this development and do not address the serious public health issue.

It is commonly described that the virus and the pandemic narrative are based on science promoted mainly by medical professionals and worldwide political and bureaucratic leadership with their suggestions and support.

"Follow the science" became the mantra for describing the existence of the novel virus (SARS-CoV-2) and related disease (COVID-19) pandemic. The general public assumes that scientists (the thinkers with vast experimentation abilities) must have done their work to make such claims.

However, who are these scientists? In general, the COVID-19 pandemic came from the description of medical experts in particular associated with world-known health authorities such as US NIH/CDC, WHO, and other similar countrywide health regulatory authorities.

[247] https://wonder.cdc.gov/vaers.html

Naturally, considering it is a health issue, it was linked to the medical profession. As a result, the medical profession declared a novel virus that caused the illness and the pandemic based on scientific studies. Further, it was stated that the needed treatment would be developing a new vaccine.

From a scientific perspective, the narrative revolves around the virus (isolation, characterization, and testing of the virus), linking the virus to disease or physiology (interaction of chemicals such as proteins and RNAs with body chemistry) and developing vaccines to kill (attacking or neutralizing) the virus. Therefore, it is essentially isolation of substances, testing, and neutralization.

Does this work reflect the training and expertise of medical professionals such as physicians?

Chapter 34—The Importance of Chemistry in Medicine

UNFORTUNATELY, MEDICAL PROFESSIONALS do not recognize that they deal with chemicals and chemistry. Hence, they make gross errors—resulting in fake viruses, invalid tests, and non-existent pandemics and treatments. One would hear non-stop advice and suggestions in the media and literature by medical experts/scientists on how medicines are developed and work. As described earlier, they are never trained as chemists or conduct research studies to develop and evaluate medicines, i.e., chemicals. Hence, making false claims causes great harm to the public and patients and, by extension, the world's economies and healthcare.

For example:

1. Believing the virus's existence.

> *SARS-CoV-2 has been sampled millions of times over from infected people, including those originally found to be infected in China.*
> —Dr Stephen Griffin, a virologist and Associate Professor at Leeds Institute of Medical Research, told Full Fact.[248]

This is a false statement. No one has seen, directly or indirectly, any specimen of the isolated and purified virus as per responses received from over 200 institutions worldwide.[249]

> *The joint research project, known as the Johns Hopkins Excellence in Pathogenesis and Immunity*

[248] https://fullfact.org/health/Covid-isolated-virus/
[249] https://www.fluoridefreepeel.ca/fois-reveal-that-health-science-institutions-around-the-world-have-no-record-of-sars-cov-2-isolation-purification/

> *Center for SARS-CoV-2 (JH-EPICS), was established under a five-year grant from the National Cancer Institute (NCI), part of the National Institutes of Health. The funding of more than $2 million per year will support studies—commencing immediately—of the immune elements that determine whether people get mild or severe COVID-19 illness following exposure to the virus.*
>
> *The center will be jointly led by Andrea Cox, M.D., Ph.D., professor of medicine at the Johns Hopkins University School of Medicine, and Sabra Klein, Ph.D., professor of molecular microbiology and immunology at the Johns Hopkins Bloomberg School of Public Health.*[250]

Note the phrase *exposure to the virus.* Such a study or project has to be fake because it would require a physical and identifiable sample of the virus, which is not available. Even vaccines for SARS-CoV-2 have been developed without using or targeting the virus but with an imaginary and non-existing virus.

A recent publication published under a variety of headings (including in news media), claiming using the virus in humans to link the disease to the virus.[251] [252] [253] [254] [255]

In short, the study describes observations after inoculating the claimed SARS-COV-2 virus to healthy human volunteers to

[250] https://www.hopkinsmedicine.org/news/newsroom/news-releases/covid-19-story-tip-new-center-deeply-explores-the-immunology-of-covid-19

[251] http://www.researchsquare.com/article/rs-1121993/v1

[252] https://www.nature.com/articles/s41591-022-01780-9

[253] https://www.nature.com/articles/s41591-022-01778-3

[254] https://www.nejm.org/doi/full/10.1056/NEJMp2106970

[255] https://www.cnn.com/2022/03/31/health/first-challenge-study-covid-19/index.html

produce the virus effects.

The study claims that volunteers were...

> *...inoculated with 10 $TCID_{50}$ of a wild-type virus (SARS-CoV-2/human/GBR/484861/2020) intranasally in an open-label, non-randomized study...*[256]

This statement is incorrect and deceitful.

The study did not use the virus but an "isolate." An isolate means part/portion of the growth culture (gunk) from a swab sample, with multiple ingredients added, including cells or debris, such as African green monkey kidney cells (or Vero cells).

The uses the word isolate in the title, not the virus, as presumed by the study's authors.[257]

It further confirms that there has never been an isolated virus (SARS-CoV-2), and no virus sample is available anywhere. Therefore, scientifically speaking, the virus (SARS-CoV-2), its illness (COVID-19), and its pandemic have to be a hoax. The publication will undoubtedly cause serious damage to the scientific credibility of the Journal and study authors. In addition, it will further expose the fraudulent aspect of virology or CDC science.

In a personal communication, the author (Saeed Qureshi) informed the Journal of the situation with a request to withdraw the publication with a false claim. The response from the Journal is awaited. Authorities can't address the issue because the issue is of science. Medical professionals and experts cannot handle the issue as they lack the needed knowledge and expertise of relevant science. Hence they started labeling such critical views and requests as spreading misinformation and conspiracy theories and enforcing censorship.

---

[256] https://www.nature.com/articles/s41591-022-01780-9

[257] https://www.ncbi.nlm.nih.gov/nuccore/OM294022.1/

1. The (PCR) test:

> *This test detects bits of the virus itself and can tell you if you're currently infected. Swabs are used to collect samples from the mucus membranes in the nose and throat where the virus may be growing or have been coughed up from the lungs. PCR tests are considered the gold-standard of NAAT testing.*[258]

Incorrect. As explained earlier, the PCR test does not test the virus or its variant, illness, infection, or COVID-19.

2. The vaccine:

> *Vaccination is one of the most effective ways to protect our families, communities, and ourselves against COVID-19.*
>
> *Evidence indicates that vaccines are very effective at preventing severe illness, hospitalization, and death from COVID-19, including against Alpha and Delta variants of concern.*[259]

Incorrect. Vaccines have never been tested against the virus in or outside the human body. More recently, the lack of vaccine efficacy has been well documented.

---

258 https://www.fredhutch.org/en/research/diseases/coronavirus/serology-testing.html

259 https://www.canada.ca/en/public-health/services/diseases/coronavirus-disease-covid-19/vaccines/effectiveness-benefits-vaccination.html

3. Clinical trials and vaccines—pseudoscience smokescreens:

> *The medical profession and its experts thrive, promoting their unique and unmatched ability to conduct clinical trials. They promote this as the most authentic and modern scientific technique, which no other medicine-related discipline has. It is the most insidious and deceptive practice and claim. As has been explained earlier, it is not a science at all. It may be considered an approach to monopolizing the medicines, their development, evaluation, indication, commercialization, and marketing.*

If clinical trial-based testing is critically evaluated, as described for pharmaceutical products earlier, it will show that it does not evaluate anything. The current use of the clinical trial approach for vaccine development may be an extreme example of the absurdity of "science" in medicine development and assessment.

To develop a vaccine (independent of formulation type, injection, or oral tablet) for treating COVID-19, one needs to get the COVID-19 patient volunteers to show that the treatment works. However, the problem is that there were no COVID-19 patients available. How could it be possible that no COVID-19 patients were available while countries were in the grip of the COVID-19 pandemic? Vaccine development exercises should have been stopped as no patients or subjects with COVID-19 were available. It is an incredible example of clinical trial fraud that authorities and experts continued developing vaccines without the availability of COVID-19 patients. How could a pandemic potentially kill hundreds of thousands of people, but one could not find an appropriate number of patients for the clinical trials?

The second possibility of conducting a trial is to inoculate, if not healthy, people, but animals with the virus to develop the illness and then treat them with the vaccine to observe its effectiveness. Again the issue was that, as described earlier, no isolated, purified, and characterized virus sample was available. So, again, it is incredible that the world is in the middle of the deadliest viral pandemic, but no one has any virus specimens. So there is a question: where are the pandemic and its virus? No one knows.

However, the authorities and experts still insisted on developing the vaccines for which patients and viruses are unavailable. So, how did the vaccine get developed?

Another overlooked aspect is how a potential vaccine candidate (irrespective of its nature, classical medicine, or mRNA-based) was chosen to conduct the clinical trial. It is impossible to select a medicine for clinical trials without knowing or having a target to treat. So, a typical and standard clinical trial/testing could not be done and has never been done.

In short, clinical trials were conducted based on a non-existing virus and non-measurable end-point (illness) with an arbitrarily selected potent chemical compound or a medicine called mRNA—promoted as a marvel wonder drug and technology. Yet, it is still a chemical compound unassociated with the body's natural composition and functioning. In addition, for testing the outcome (efficacy), a random and scientifically invalid test, known as PCR, was used. How did such a thing happen and not get caught in an audit? Unfortunately, in the medicines (development and testing) areas, an audit is conducted by peers, not by independent third-party auditors, who are taught and practice following the narrative, considered the "science."

Therefore, made-up clinical trials were conducted using healthy human volunteers, half injected with vaccines (an arbitrarily selected chemical compound called mRNA) and half without vaccines. For example, the following description and interpretation is based on documentation and data submitted to the FDA by Pfizer for its mRNA vaccine development—the first and perhaps most prescribed one.[260]

Volunteers in both groups were monitored with the PCR test to observe COVID-19 in healthy volunteers. It is important to note that this PCR test has never been validated to test the presence or absence of the COVID-19 illness or its virus. However, a positive test is considered to indicate the presence of the virus. It is still a mystery how a non-validated PCR test becomes a scientific and acceptable marker for the virus or the illness. The only reason is that everything is reviewed and assessed by peers or expert committees involving the peers—a cult-like situation.

In Pfizer-BioNTech vaccine development, 162 volunteers tested PCR positive for placebo while only eight were in the vaccinated group, out of about twenty thousand healthy volunteers in each group. It indicates an overall positivity of 0.81% (falsely considered COVID-19 infection because PCR test never determines virus and/or its infection). It is assumed that if the group would not have vaccinated, then it would have produced 182 positives as well; hence vaccination "cured" the 95% [(162-8)/162)*100)=95%], i.e., vaccine declared as 95% efficacious.[261]

The basis of this interpretation is as follows: if there had not been given the treatment, both (treated and placebo) groups would have an equal number of positive tests. However, as the treated group showed only eight positives, the vaccination protected the remaining from the virus. Make sure it should be understood that the test does not test the virus at all. However, claims are being made about the protection from the virus. How could such an interpretation be considered something other than deception? Unfortunately, the implied conclusion that the public generally assumes that out of twenty thousand plus volunteers in the

[260] https://www.fda.gov/advisory-committees/advisory-committee-calendar/vaccines-and-related-biological-products-advisory-committee-december-10-2020-meeting-announcement

[261] https://bioanalyticx.com/the-fda-committees-review-of-pfizer-biontech-covid-19-vaccine-unscientific-false-and-deceitful/

treatment group, 95% (19 900) must have been cured of the illness is incorrect. It is a sad and disturbing story of false science and begs for answers, clarification, and an audit of the fake "science of medicines (AKA CDC-science)."

This is the continuation of negligence and misrepresentation of science in medicine. The development of vaccines provides the latest and extreme example of non-scientific but fraudulent practices in medicine. Most clinical trials would fail to meet the fundamental principle and scientific requirements if critically evaluated, even superficially.

One thing to note is that clinical trials are not scientific studies in general. Instead, they are like a survey conducted for social or marketing observations regarding the usefulness of some items. For example, consider if McDonald's restaurant likes to introduce a new burger. The company will provide the volunteers with a new burger and assess the reactions, likes, or dislikes. If the majority liked the burger as the company hoped for, it would be marketed to the general population. Otherwise, the study would be wrapped up.

Similarly, a clinical trial is, in general, a survey-based exercise. Healthy volunteers (or, in some cases, patients) are given medicine (against no medicines, i.e., placebo), and reactions (based on symptoms of illness and/or testing) are noted that decide the usefulness of the medicines.

The actual scientific part of medicines testing is done in animals or in vitro with analytical chemical testing, i.e., developing tests for the illness or its markers and selecting appropriate medicines.

4. The suggested mechanism of vaccination action is often depicted, as shown in Figure 41. Although such is

shown in the scientific literature,[262] it is, unfortunately, an incorrect, representation of the mode of action. A vaccine or vaccination never kills the virus in this manner or at least directly.

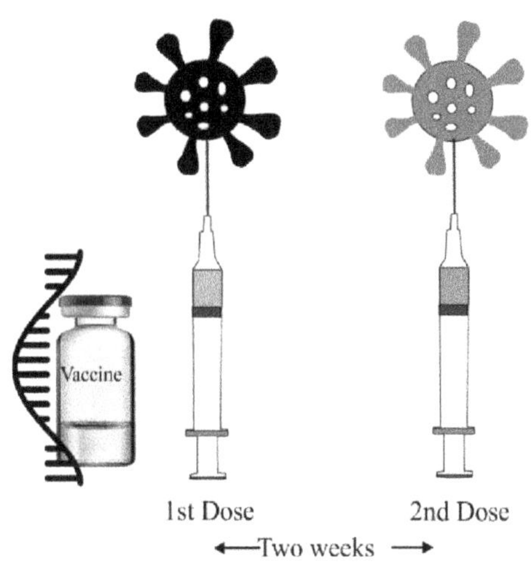

Figure 41
Schematic representation of vaccination against the virus (SARS-CoV-2)

A more representative mode of action for vaccines is shown below (Figure 42). In this case, mRNA is enclosed in a lipid nanoparticle (more accurately, droplets) and is injected into the body. These droplets go into the body and release the mRNA molecules, which presumably act as a blueprint for manufacturing virus spike proteins. The body assumes this (protein) as a foreign body and triggers its defense mechanism by producing antibodies (specific types of proteins). In the future, if the same virus comes in, the body will recognize it with its stored capacity to generate antibodies and potentially kill the spike protein and its virus. Therefore,

[262] https://theconversation.com/how-mrna-vaccines-from-pfizer-and-moderna-work-why-theyre-a-breakthrough-and-why-they-need-to-be-kept-so-cold-150238

vaccines do not cure a person's illness, infection, or death but would kill the virus only before they cause the infection. It is assumed that the injected mRNA produces specific proteins (spike proteins) of the SARS-CoV-2 virus.

However, no isolated spike protein sample is available because no isolated virus has been extracted for the isolated spike protein.

So, how did this theory or mechanism of action has been proposed?

Pure and simple imagination and illusion which is considered science in medical professions. It is precisely like the situation described earlier in pharmaceutical product development. It all falls under the chemistry subject described earlier, i.e., isolation, purification, and characterization for both the virus and mRNA technology.

But, unfortunately, nothing of this scientific part has been done. This is a regrettable situation that science is going through at present. The literature does not provide any scientific evidence that this proposed mechanism of action is studied and valid. It is just the pandemic story that the virus causes disease. There is no specific disease with any particular symptoms, and no virus has ever been isolated to link to the disease.

Similarly, vaccines have been developed to produce proteins; no one knows the exact nature of the proteins and their benefits and adverse effects. However, the marketing has been done as if the mRNA is pure and can produce only one specific type of protein. Nothing could be far from the truth. It is highly unlikely that mRNA would be pure—as often the case for chemical or biological processes.

A pure specimen of SARS-CoV-2 spike protein has never been obtained; therefore, it would be correct to assume that mRNA capability to produce spike protein has not been established.

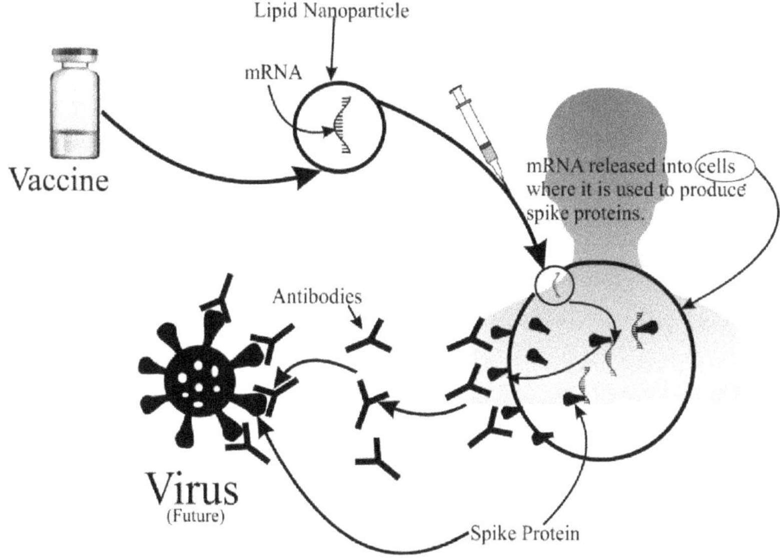

Figure 42
Schematic Representation of the Mode of Action of an mRNA-Based Vaccine

Considering the above-described facts, i.e., lack of scientific basis for vaccine development, it is impossible to develop an appropriate and valid vaccine as predicted almost a year ago.[263]

Here are more examples of the nonstandard science practices in the medicines area:

1. Experiments of the so-called virus's presence in patients have been reported without any parallel control, blank, placebo (an invalid science practice).
2. Conclusions are based on a single patient/observation (i.e., n = 1). Anyone with a basic understanding of science and experimentation would reject such data without fail.

[263] https://bioanalyticx.com/science-for-the-pandemic-at-the-authorities-false-in-fact-fraudulent-requires-urgent-action/

3. Epidemiologists ignore such a weakness (n=1) for their modeling and projections. Further, it is unclear how they obtained the variance values to simulate and project the virus's spread and transmission.
4. During the pandemic, a famous phrase was "flatten the curve," meaning controlling the spread and intensity of "cases" in a given population. For example:

> *If you look at the curves of outbreaks, they go big peaks, and then come down. What we need to do is flatten that down, Anthony Fauci, director of the National Institute of Allergy and Infectious Diseases, told reporters Tuesday.*
>
> *That would have less people infected. That would ultimately have less deaths. You do that by trying to interfere with the natural flow of the outbreak.* [264]

Surprisingly, no baseline curve/chart has been provided. How should one know a population's normal (PCR) response curve? A scientific study requires some reference points/curves but missing.

5. No specimen is available for the virus or its variant. However, medical experts claim the virus has been isolated and available—this view reflects following instructions as trained rather than critically evaluating scientific requirements and facts.
6. Scientifically, no test can be developed and validated without the availability of a reference standard, in this case, an isolated and purified virus or RNA specimen. Hence, the claims and recommendations of the PCR, Rapid/Antigen tests reflect incompetency or scientific

[264] https://www.statnews.com/2020/03/11/flattening-curve-coronavirus/

fraud.

7. To treat an illness, first, the illness has to exist, e.g., in the case of COVID-19, which is allegedly a disease caused by a virus called SARS-CoV-2. However, no one has or seen the virus in any patient; hence, there is no reason to believe that virus exists. Therefore, the associated disease COVID-19 cannot exist either. Unfortunately, medical experts, particularly physicians, mistakenly believe in the existence of the virus and the disease.[265]
8. The diagnosis is mainly based on the irrelevant and invalid PCR test. Therefore, the disease (COVID-19) is undoubtedly a false diagnosis. But, as some say, why are people get sick then? Firstly, the illness appears infectious (parasitic?), not COVID-19, and treatable with anti-infection medications, as noted earlier, with all patients claimed to have COVID-19 being treated successfully with antibiotics.

Secondly, ivermectin appears to be a good match for the infection, as Dr. Kory, in one of his published articles, suggests.[266]

> *It means there would not have a pandemic if physicians applied their learned knowledge and training (diagnosing and treating with recommended treatments), as Dr. Kory suggested.*

9. "President Donald Trump on Monday announced a strict set of guidelines for Americans to follow for the next 15 days to try to "slow the spread" of the coronavirus. The

[265] https://bioanalyticx.com/video-virus-covid-pandemic-vaccine-and-testing-fiction-not-reality-or-science/

[266] https://www.ncbi.nlm.nih.gov/pmc/articles/PMC8088823/pdf/ajt-28-e299.pdf

> recommendations call on people to sharply limit their normal behaviors when it comes to eating out, shopping and socializing." (MAR 16 2020).[267]

This is an opinion that lacks scientific/experimental support.

10. CDC calls on Americans to wear masks to prevent COVID-19 spread.[268]

The CDC provided the following supporting evidence for the efficacy of the mask use (note earlier, CDC categorically declined the relevance and efficacy of the masks).

> *There is increasing evidence that cloth face coverings help prevent people who have COVID-19 from spreading the virus to others.*
>
> *The results of the Missouri case study provide further evidence on the benefits of wearing a cloth face covering.*
>
> *The investigators found that none of the stylists' 139 clients or secondary contacts became ill, and all 67 clients who volunteered to be tested showed no sign of infection.*

11. The second study published in JAMA provided further evidence, in experts' view, in support of masks used, describing, as per CDC science:

Universal Masking to Prevent SARS-CoV-2 Transmission—The Time Is Now

The authors present data that prior to the

[267] https://www.cnbc.com/2020/03/16/trumps-coronavirus-guidelines-for-next-15-days-to-slow-pandemic.html

[268] https://www.cdc.gov/media/releases/2020/p0714-americans-to-wear-masks.html

> *implementation of universal masking in late March 2020, new infections among HCWs with direct or indirect patient contact were increasing exponentially, from 0% to 21.3% (a mean increase of 1.16% per day). However, after the universal masking policy was in place, the proportion of symptomatic HCWs with positive test results steadily declined, from 14.7% to 11.5% (a mean decrease of 0.49% per day). Although not a randomized clinical trial, this study provides critically important data to emphasize that masking helps prevent transmission of SARS-CoV-2.*[269]

The examples provided above (# 9 and #10) formed the "scientific" basis of masks' use and efficacy. However, as noted in example (#10), such would not qualify as a scientific study or evidence but perceived and implied as scientific but are anecdotal observations. Such claims, in reality, show a pathetic level of incompetency in science by the authorities and experts.

12. Consider evaluating the use of masks as a protective barrier against the virus. As described earlier, this can be established with a simple experiment, i.e., passing the virus through the mask or its material and measuring restrictive ability.

In reality, no worker should be working in a laboratory without an assurance that protective gear will protect them from the virus. But unfortunately, no such protection or assurance is available, just word of mouth.

When some were questioned during the SARS-CoV-2 pandemic to show the effectiveness of the masks' use, suddenly

---

[269] https://jamanetwork.com/journals/jama/fullarticle/2768532

studies started appearing in the literature from medical institutions. An example of one of many such studies is provided here.[270]

Describing such a study:

> *Guidelines from the CDC and the WHO recommend the wearing of face masks to prevent the spread of coronavirus (CoV) disease 2019 (COVID-19); however, the protective efficiency of such masks against airborne transmission of infectious severe acute respiratory syndrome CoV-2 (SARS-CoV-2) droplets/aerosols is unknown.*
>
> *Here, we developed an airborne transmission simulator of infectious SARS-CoV-2—containing droplets/aerosols produced by human respiration and coughs and assessed the transmissibility of the infectious droplets/aerosols and the ability of various types of face masks to block the transmission. We found that cotton masks, surgical masks, and N95 masks all have a protective effect with respect to the transmission of infective droplets/aerosols of SARS-CoV-2 and that the protective efficiency was higher when masks were worn by a virus spreader.*
>
> *Importantly, medical masks (surgical masks and even N95 masks) were not able to completely block the transmission of virus droplets/aerosols even when completely sealed.*

Note the wording in the last line above, *virus droplets/aerosols.*

There is no such thing as "virus droplets/aerosols." Experiments were conducted using only the "droplets/aerosols," but the conclusion is drawn to reflect the virus particle. It is unclear

---

[270] https://journals.asm.org/doi/10.1128/msphere.00637-20?permanently=true

how such a study would be considered a scientific study—unfortunately, it is not.

13. *This systematic review and meta-analysis are designed to determine whether there is empirical evidence to support the belief that "lockdowns" reduce COVID-19 mortality. Lockdowns are defined as the imposition of at least one compulsory, non-pharmaceutical intervention (NPI). NPIs are any government mandate that directly restrict peoples' possibilities, such as policies that limit internal movement, close schools and businesses, and ban international travel.*

    An analysis of each of these three groups support the conclusion that lockdowns have had little to no effect on COVID-19 mortality. More specifically, stringency index studies find that lockdowns in Europe and the United States only reduced COVID-19 mortality by 0.2% on average. SIPOs [shelter-in-placeorder] *were also ineffective, only reducing COVID-19 mortality by 2.9% on average. Specific NPI studies also find no broad-based evidence of noticeable effects on COVID-19 mortality.*

    In consequence, lockdown policies are ill-founded and should be rejected as a pandemic policy instrument.[271]

14. *Data from South Africa and the United Kingdom demonstrate that vaccine effectiveness against*

---

[271] https://sites.krieger.jhu.edu/iae/files/2022/01/A-Literature-Review-and-Meta-Analysis-of-the-Effects-of-Lockdowns-on-COVID-19-Mortality.pdf

> *infection for two doses of an mRNA vaccine is approximately 35%.*[272]

Rather than reconsidering the situation and correcting it, a play of new variants (Delta and Omicron) was initiated. There is no valid test method available so far for the original virus; one wonders how new variants have been detected. They cannot be. The medical and pharmaceutical experts and authorities' "science" is based on narrative, accepted, and approved by peers. Hence, it is accepted that there are new variants opined by the same peers.

It is now a well-known fact that vaccines have not been effective as predicted.[273]

The reason is that logically and scientifically it is impossible to develop a treatment for something (virus, protein, or RNA) when its reference sample is unavailable. Therefore, a new illusionary story has to be developed to rationalize the use of fake vaccines.

The lack of efficacy has been acknowledged indirectly, stating that they are breakthrough positive cases due to variants. However, claims have been made that vaccines cause less serve diseases.[274]

> *Omicron spreads more easily than the original virus that causes COVID-19 and the delta variant. However, omicron appears to cause less severe disease. People who are fully vaccinated can get breakthrough infections and spread the virus to others. But the COVID-19 vaccines are effective at preventing severe illness.*

---

[272] https://www.cdc.gov/media/releases/2021/s1227-isolation-quarantine-guidance.html

[273] https://spectator.com.au/2022/07/vanishing-vaccine-mandates/?fbclid=IwAR3SGMsx7jj67mRxlrQ1_s4gx-aRVBZI-7zvj0KwQJRcpdVeCyOAYpIr-OE

[274] https://www.mayoclinic.org/diseases-conditions/coronavirus/in-depth/coronavirus-vaccine/art-20484859#covid-variants

> *In April, the CDC downgraded the delta variant from a variant of concern to a variant being monitored. This means that the delta variant isn't currently considered a major public health threat in the U.S.*
>
> *Vaccines reduce the risk of COVID-19, including the risk of severe illness and death among people who are fully vaccinated.*[275]

The problem is that the public assumes that these claims must be based on scientific studies. However, it is not true. No one conducts scientific studies, and they cannot be conducted because such studies require a specimen of the virus and a valid test for the virus or its infection. Both do not exist. Such conclusions are based on surveillance studies using PCR test outcomes, i.e., counting the numbers with fake test. Therefore, findings have to be fake and irrelevant. Under one of the subheadings, COVID-19 Vaccine Effectiveness Research, it is described that:

> *Clinical trials are conducted to determine vaccine efficacy before the U.S Food and Drug Administration (FDA) determines whether to approve a vaccine. CDC and other partners assess COVID-19 vaccine effectiveness under real-world conditions after the FDA approves a vaccine.*
>
> *Below you'll find descriptions of current and planned vaccine effectiveness evaluations CDC is conducting with partners. The descriptions include the evaluation's data collection platform, protocol (if available), outcome, population, and*

[275] https://www.cdc.gov/coronavirus/2019-ncov/vaccines/effectiveness/work.html

> *participating sites.*[276]

It is important to note that clinical trials were conducted to evaluate the efficacy of the vaccines based on the symptoms and/or positive PCR test results. Hospitalization and deaths were not the criteria for assessing the vaccine's efficacy in clinical trials. Therefore, claiming effectiveness with criteria of hospitalization and deaths is a misrepresentation of data or reports. Such claims should be withdrawn. Not only have the vaccines not been found effective, but their safety or toxicity record has been dreadful.

15. First, there have not been any relevant and valid toxicity studies reported using vaccines in humans. Therefore, it is scientifically deceitful and immoral to state that COVID-19 vaccines are highly safe. It is simply factually false to make such claims. Animal studies provide safety and toxicity data and clues about what to look for in humans, which have not been done. It has been stated that animal toxicity studies are underway and will be completed shortly. However, it is a lie. Scientifically valid studies can never be done because one would require a pure isolated virus sample and a valid test for such studies. Unfortunately, both are missing; hence studies cannot be conducted.

It is the biggest irony that physicians gained high respect in the eyes of the public by providing needed and valuable services. Unfortunately, however, they have lost it all by indulging themselves in the activities (science and research) they are never trained for. In actuality, medical experts have committed a science fraud and caused enormous harm to the people and the countries. They may try to recover the lost respect and credibility by helping

---

[276] https://www.cdc.gov/vaccines/covid-19/effectiveness-research/protocols.html

the public/patients with the expertise and knowledge they have been trained for. However, at present, they may not be able to. They require independence to work with patients to help and prescribe medicines (chemicals) from independently manufacturing facilities and not only governments' "approved" chemical manufacturing facilities. When one looks closely at drug treatments, developments, administration, discussions, and choices are made between the medical profession, pharmaceutical industry, and the authorities such as FDA and CDC. Medications are being forced onto the patients, in fact, healthy people, without their consent and input. Physicians-patients interaction is almost non-existent in this respect.

Consider the example of the coronavirus pandemic. Healthy people are being forced (mandated) with a government-imposed disease and medication. Patients and many physicians are desperately requesting options for reconsideration, but only the government/bureaucracy has been deciding for them. Governments have accepted this view that all citizens are sick, or going to be sick, with some mysterious disease. Hence, they have to be treated.

It is time to think clearly and differently about how medical and pharmaceutical-based professionals have deviated from their mandate and training. They have indulged in a subject they have never been trained in. As a result, they make colossal mistakes and do not look caring and smart in the public's eyes except in the eyes of "experts" and peers.

They have committed a colossal fraud.

One of the options to address the situation is to consider separating the development and manufacturing of medicines from the medical profession. Allopathic medicines, in particular, are chemicals and should be part of chemical manufacturing and development. In addition, considering the medicines development, testing, manufacturing, and sale in the medical and pharmaceutical profession have created a massive conflict of interest situation. The situation is that the medical professionals are acting as sale agents for vaccination (or pharmaceuticals).

Anyone who likes to live their normal life must have the vaccine (worldwide) even when clearly shown that vaccines are not effective at all, and causes potentially severe adverse effects, including deaths.

The question is, why so many unsupported scientific claims are being made?

The reason is that medical and pharmaceutical are not scientific disciplines but service-based, using science for their services. It is like a chef cooking and serving a meal per the customer's need using well-established agricultural or food products. However, chefs are never trained to grow, develop, characterize, test, or introduce new varieties of crops. Similarly, physicians, pharmacists, and related experts have never been trained to develop, test, and characterize new diseases and medicines but are trained to use their knowledge of existing well-established diseases and drugs to provide service to clients, i.e., patients.

A car mechanic never develops a car battery, tire, engine, etc., but one always goes to them for change or repair. It could be argued that with the lack of this understanding, medical professionals have ruined their profession by falsely considering themselves trained scientists or claiming that they "follow the science." Not only has this practice of illogical and unscientific clinical trials ruined the development of proper allopathic medication, but it has also severely hindered other useful medicines such as nutritional and herbal based. Unfortunately, alternate medicine professionals are also equally unaware of critical deficiencies of the clinical trial and downplay some clear advantages of their treatments.

Some other thought and mind-twisting tricks and discussions, in the name of science, to keep the public looking the other way, such as establishing the origin of the virus (SARS-CoV-2), the gain of function research, debating germ vs terrain theory (aka vaxers vs anti-vaxxers).

Concerning the debate about the origin of the virus, this article summarizes the issue very well.

> *Since the first reports of a novel severe acute respiratory syndrome (SARS)-like coronavirus in December 2019 in Wuhan, China, there has been intense interest in understanding how severe acute respiratory syndrome coronavirus 2 (SARS-CoV-2) emerged in the human population.*
>
> *Recent debate has coalesced around two competing ideas: a "laboratory escape" scenario and zoonotic emergence. Here, we critically review the current scientific evidence that may help clarify the origin of SARS-CoV-2." Twenty-three (23) authored the publication on the subject of the virus presumed to be a lifeless particle but potentially can grow and multiply, like a seed, under a specific environment. Most of the authors are from biology, virology, or microbiology, with a couple from medicine-related areas. Should it not be interesting to note that none of the authors belongs to material science, considering the virus is presumed to be a particle made of well-established chemical molecules (RNA and proteins)? Moreover, such molecules could appropriately be characterized in chemistry-based laboratories.*
>
> *So, the obvious question is, why is it so? The simple reason is that a material (chemistry)-based scientist will not accept the virus arguments. It is because, before seeking the origin, the scientist will ask for a complete physical description of the virus and to have its specimen. However, the problem is that the actual description and specimen of the virus do not exist. Therefore, the*

> *story of the virus will collapse immediately. The direct and logical way to resolve this issue would be to have the virus from the environment that infects people, i.e., isolate it from people or patients. Then compare it with the virus, stored specimen (modified or unmodified). However, the problem is that no one has the virus specimen from any source. Therefore, it is impossible to identify or compare the virus to establish its origin or anything else. Therefore, discussion of the origin of the virus continues without the involvement of actual scientists with relevant and needed expertise—hoping that one day somehow, the virus will appear from somewhere (human, monkey, bat, cow, chicken, etc.). Till such time search and discussion should continue—unfortunately, it is a discussion for the sake of discussion! The fact remains that as there is no evidence of the virus's existence, its origin or source cannot be determined.*[277]

It is a psychological game to keep discussing an imaginary or fictitious object to convince that virus is real, like "fake it until make it."

Therefore, the discussion of the origin of the virus has no relevance or benefit to science or public health. Similarly, the gain of function (GoF) discussions and inquiries are simply distractions from the practice of fraudulent science. In simple terms, GoF means modifying the original product or process to improve its functionality. If the product and process have deleterious properties, those properties could also be enhanced under GoF. In

---

277
https://www.sciencedirect.com/science/article/pii/S0092867421009910

principle, GoF itself is not a bad concept. However, most innovation and development may be considered beneficial and sought after, with restriction policies for controlling and prohibiting dangerous outcomes. In the case of virus research, it is assumed that GoF is about making a virus from dangerous to deadlier.

However, it may not be an accurate assumption. A more disturbing aspect of GoF concerning the virus is that it is a lie. No one has worked on the virus because it never existed. Most of the work reported or discussed is related to computer-generated RNA, which has no relevance to the virus or its properties.

No one has seen an actual specimen of isolated RNA from the virus, so then how could its structures be modified or improved? It cannot be. Discussing computer-generated arbitrary long chain structures has no relevance or benefit to science or the public's health. It would be much more beneficial if the time and financial resources were spent assessing the regular ailments with simple and classical treatment.

Chapter 35—Germ vs Terrain Theory

SOME THOUGHTS ON germ vs terrain theory: talking about illness and virus would not be complete without saying a few words about the so-called germ and terrain theories.

The germ theory proposes that certain diseases are caused by microorganisms' invasion of the body, organisms too small to be seen except through a microscope. The French chemist and microbiologist Louis Pasteur, the English surgeon Joseph Lister, and the German physician Robert Koch are credited for developing and accepting the theory. In the mid-19th century, Pasteur showed that organisms in the air cause fermentation and putrefaction.

In the 1860s, Lister revolutionized surgical practice by utilizing carbolic acid (phenol) to exclude atmospheric germs and thus prevent putrefaction in compound fractures of bones. In the 1880s, Koch identified the organisms that cause tuberculosis and cholera.[278]

Terrain theory argues that if the body is well and balanced, germs that are a natural part of life and the environment will be dealt with by the body without causing sickness. The concept of virus and virology appears to have come from the germ theory, i.e., virus (microorganism) being germ. People who support the virus or virology provide pictures of the virus, such as SARS-CoV-2, indicating the germs are present; hence people are sick with them.

In this regard, the first thing to consider is that these are theories, not scientific and proven correct, at least yet. Showing the presence of some particles with the help of a very powerful camera (AKA electron microscope) does not mean that the presumed particles are viruses and cause illness or death. It is like finding some unusual amount of cash in someone's wallet does not prove a

[278] https://www.britannica.com/science/germ-theory

person is rich or the money is obtained by illegal means, and by extension, the person is a criminal. The court system is there to evaluate the legitimacy of the money and its possession.

Similarly, one must prove that the observed particles are real and indeed caused the illness; this is the science part (the court/justice system) that comes into play. Science does not mean getting or promoting an opinion or view of a person or organization that carries the word "science" in its name, academic degree, or description. Science means conducting and showing some physical procedures (experiments) supporting the claims. Unfortunately, this is exactly the situation in the medical science area for the past few decades, particularly during the past two years of dealing with the COVID-19 pandemic.

As an example, in a US Senate hearing, Dr. Fauci, director of the National Institute of Allergy and Infectious Diseases (NIAID) at the National Institutes of Health (NIH), fired back against criticism by Republican lawmakers insisting his guidance throughout the coronavirus pandemic has been based on scientific evidence and that "science and truth are being attacked." Republican lawmakers have accused Dr. Fauci of flip-flopping on wearing masks and downplaying the possibility the pandemic started with a lab leak in Wuhan, China.

His coronavirus guidance has evolved with new data: "That's what's called…the scientific process," Dr. Fauci said. "As you get more information, it's essential that you change your opinion because you've got to be guided by the science and the current data." While Dr. Fauci initially dismissed the idea of wearing face masks to prevent the spread of the virus in early March 2020, he changed his mind after new evidence emerged that masking significantly cut down on transmission, along with the news that up to half of the new cases at one point in the pandemic was passed along by an asymptomatic carrier, Dr. Fauci said Wednesday. Fauci expressed concern that political attacks will discredit him and other health officials and hurt their credibility with the public when they issue public health recommendations.

"People want to fire me or put me in jail for what I've done," Dr. Fauci said. "Namely, **follow the science.**" [279] [Emphasis added]

The question is: why is he saying that he followed the science?

Why not show the experimental procedures and details to the people and let the science speaks for itself whether it has been pursued?

The mask's effectiveness in protecting one from the virus is a case in point. Rather than stating the following-the-science mantra, he should have submitted a report from an experiment showing a mask or its material protecting or absorbing the virus resulting in assumed protection.

Scientifically speaking, the usefulness of wearing face masks can be established with a straightforward laboratory experiment, as described earlier. However, experiments would require the virus and a method to monitor the virus. Experts ("scientists") do not tell the public that both, i.e., actual virus and a valid test to monitor the virus, are not available, which will collapse the whole virus and pandemic fraud.

Clearly, the science has not been followed.

So, why does Dr. Fauci says science has been followed? It is because he has been labeled as a scientist and is the part/head of the organization as a world-known "science" organization. Therefore, whatever that person's opinion is, or the organization's, should be taken as science.

This lack of science issue is not limited to Dr. Fauci or his organization, but worldwide because they all have been taught to say and preach the same thing in a cult-like manner and declare one another the science experts. In layman's terms, they are effectively lying about science like a con artist and are not scientists by any criteria or definition of science. However, people often ask how one would say such a thing about a noble profession and its

---

[279] https://www.forbes.com/sites/carlieporterfield/2021/06/09/fauci-on-gop-criticism-attacks-on-me-quite-frankly-are-attacks-on-science/?sh=15a372e34542

professionals.

Although, looking closely, these people are primarily medical doctors or physicians. No doubt, they are trained to provide medical services to people. However, they have never been taught or trained as scientists. Instead, they are trained as repairpersons, not the scientists as explained in detail previously under section science in the medical profession.

So, it is hoped that people will understand that germ theory is a theory and is not a scientifically proven fact showing the cause of the illness or death. In reality, germs are needed to link, experimentally, to the body and the cause of the disease. For example, suppose someone gets buried in the mud and dies, but it does not mean the person died because of the mud.

On the contrary, the person might have died of suffocation because of the covering with the mud. Therefore, superficial observation does not lead to explaining the disease. People do not realize that presence of germs alone cannot explain the cause of illness. One needs to explain how germs change the normal working mechanism of the body, the chemical environment, and the processes that cause illness or death. So the causes of the diseases have still to be found scientifically. That can only be established by studying the science, i.e., body physiology and chemistry, following fundamental science principles, as explained earlier.

On the other extreme of germ theory is terrain theory. Terrain theory argues that if the body is well and balanced, germs that are a natural part of life and the environment will be dealt with without causing sickness. "Germs are a natural part of life," that is, they are not harmful. It argues that the more important part is working on the 'terrain,' the body's inner environment making it inhospitable to viruses, parasites, etc. It is not clear how this theory came to such a conclusion. The body is not immortal. It has to die of its natural cause, usually indicated by illness.

So, where does disease come from, and what is a condition?

Arguably, this theory does not define health or a healthy body and how the stress-free, good food environment interacts with the healthy body.

Where are experimental details to observe or describe the details? To prove this theory's validity, one must first understand the body's baseline mechanics, i.e., physiology and chemistry. It is missing; this is a science part that needs to be studied.

Unfortunately, the public remains in the hands of self-proclaimed scientists, AKA con artists; hence, true science is currently lacking.

Chapter 36—The Bernician: Defendants & Their Gates Connections

IN A SUPERB online post, *The Defendants & Their Gates Connections*, The Bernician offers a superb outline of criminal conspiracy among the British political elite.[280]

He writes in a series of posts exposing the multitude of frauds, of which Matt Hancock, Chris Whitty, Patrick Vallance and Neil Ferguson are prime suspects.

Chris Whitty

The Bernician writes of Whitty as the second defendant in a criminal Covid prosecution:

> *For the purposes of understanding the critical part the 2nd defendant has played in perpetrating the crimes alleged, it is somewhat illuminating to consider the role Bill Gates played in his professional career, before he became the UK Government's Chief Medical Officer in late 2019.*

The Bernician's post shows that:

> *In short, the 2nd defendant is highly qualified in medicine, medical law and economics, so he has no excuse for not knowing that:*
>
> *a. The COVID treatments granted emergency licenses by the MHRA are not 'vaccines', under the long established medical and*

[280] https://www.thebernician.net/author/mob/

legal definitions, given that it has been widely reported that 'vaccination' does not prevent people from contracting or passing on the 'virus', thereby entirely negating it as a preventative treatment.

b. The administration of the UK Government approved untested COVID vaccines is a breach of the Nuremberg Code, on the ground that it is tantamount to involuntary medical experimentation.

c. The consequences of the lockdown policies imposed would be catastrophic for civil liberties, the 'National Debt' burden on the taxpayer and the nation's solvency.

d. The results of the flu and COVID 'vaccine' safety studies emphatically showed that there were and still are potentially debilitating and fatal side effects.[281]

LSHTM & The Gates Foundation

The Bernician then details how, in 2008, the London School of Hygiene and Tropical Medicine [LSHTM] was awarded grant funding of over $46.4 million by the Gates Foundation and $12.7 million from other partners for research into treatment and prevention of malaria, tuberculosis and HIV/AIDS.

The largest portion of that funding went to the ACT Consortium, a body of research institutions around the world that conducted research into treatment for malaria and of which the LSHTM was a member. The ACT Consortium received around $40 million of the funding.

At the time, the 2nd defendant (Whitty) was a director and Principal Investigator of the ACT Consortium. According to

[281] https://edition.cnn.com/2021/01/08/health/covid-vaccinated-infected-wellness/index.html

reports from the Gates Foundation, the money went via the LSHTM, which Whitty represented at a senior level and took material benefit from upon its receipt, since it funded his research into malaria.

This research grant was an integral part of the Gates Foundation's drive to maximise the 'vaccination' uptake in Africa and the rest of the world, which developed into the 'Decade of Vaccines' project, from 2010-2020. The project was singularly dedicated to maximising 'vaccination' uptake worldwide over the course of a ten-year plan, in which Whitty was engaged to play his integral part.

This established a financial motivation connecting Whitty to Gates and implicates them in a conspiracy. A year after accepting the £31 million grant from Gates, Whitty was appointed Chief Scientific Adviser and director of research for the Department for International Development (DFID); and that he led the Research and Evidence Division, which worked on health, agriculture, climate change, energy, infrastructure, economic and governance research.

Also, not widely known is that Whitty serves as an Executive Board Member of the World Health Organization (WHO). Such WHO board members decide the agenda for the World Health Assembly and the resolutions to be considered by the Health Assembly. The Bernician explains that Whitty began his three-year term as an Executive Board member on January 01, 2020, two weeks after the board recommended the Global 'Vaccine' Action Plan to the Health Assembly, in which it was declared:

> *The Strategic Advisory Group of Experts on immunization, at its meeting in October 2019, proposed that a post-2020 immunization strategy should:*
>
> *(1) ensure more timely and comprehensive implementation at global, regional, national and subnational levels;*

(2) focus on countries, specifically:—place countries at the centre of strategy development and implementation to ensure context specificity and relevance;—strengthen country-led evidence-based decision-making;—encourage the sourcing and sharing of innovations to improve programme performance;—promote use of research by countries to accelerate uptake of vaccines and vaccine technologies and to improve immunization programme performance.

(3) maintain the momentum toward the goals of the global vaccine action plan:—incorporate key elements of the global vaccine action plan, recognizing its comprehensiveness and the importance of sustaining successes in;—immunization every year; add a specific focus on humanitarian emergencies, displacement and migration, and chronic political and socio-economic fragility;—encourage stronger integration between disease-elimination initiatives and national immunization programmes;—encourage greater collaboration and integration within and outside the health sector.

(4) establish a governance model better able to turn strategy into action:—create a robust and flexible governance structure and operational model based on closer collaboration between partners;—incorporate the flexibility to detect and respond to emerging issues;—develop and maintain a strong communications and advocacy strategy.

(5) promote long-term planning for the development and implementation of novel vaccine and other preventive innovations, to ensure that populations benefit as rapidly as possible;

> *(6) promote the use of data to stimulate and guide action and to inform decision-making;*
>
> *(7) strengthen monitoring and evaluation at the national and subnational levels in order to promote greater accountability.*

It is beyond doubt that Whitty was serving the Bill Gates agenda to maximise 'vaccination' uptake worldwide, by and through establishing 'safe markets' for 'vaccines', via the criminal monopolisation of international government policy.

But, perhaps a more damning fact is that Whitty has been integral in the formation of CEPI. CEPI is the Coalition for Epidemic Preparedness Innovations, which was launched at Davos in 2017 by the governments of Norway and India, the Gates Foundation, the British-based Wellcome Trust 'global health charity' and the World Economic Forum.

CEPI state on their website that they are:

> *...working together to accelerate the development of vaccines against emerging infectious diseases and enable equitable access to these vaccines for people during outbreaks.*[282]

Whitty served on the interim board of CETI and it cannot therefore be doubted he holds a vested interests in Gates' plan to maximise vaccination uptake worldwide as his memberships of the UK Government's 'Vaccine' Taskforce, the UK 'Vaccine' Network and the WHO's Executive Board emphatically affirm.

CEPI & The UK 'Vaccine' Taskforce

On April 17, 2020, it was announced in a press release that the UK Government had granted £250 million of taxpayer's money to

[282] https://cepi.net/about/whyweexist/

CEPI. This coincided with the launch of the UK's 'Vaccine' Taskforce, of which Whitty was named as a member.

Whitty thereafter gave a press release stating:

> *The UK has some of the best vaccine scientists in the world, but we need to take account of the whole development process. This taskforce will ensure the UK can take an end-to-end view. This includes funding research, like the recent NIHR/UKRI call, and ensuring manufacturing capability to deliver a COVID-19 vaccination as quickly as possible.*

In effect, Whitty was a key player in orchestrating the advance of Imperial College, UCL, LSHTM, Sheffield University and Oxford University researching 'vaccines' and taking control of every part of the development process. The outcome was the creation of a COVID 19 'vaccine' monopoly, controlled by Bill Gates and his partners and paid for by the UK taxpayer.

Moreover, by Whitty's hand he could enforce the indemnification of both manufacturers and those who administer the jabs, ensuring the 'vaccines' they developed could skip the pre-requisite decade of safety tests and be offered as soon as possible to everybody in the UK and then the rest of the world.

National Influenza Immunization Program

The Bernician's post then tells us how a expanded National Influenza Immunisation Program came about because Whitty signed off on a letter to every GP practice in the UK on August 05, 2020, launching a full scale aggressive strategy to maximise flu 'vaccination' uptake, all developed by Gates funded Sheffield University in 2011.

From the above, it is clear that the purpose of the marketing campaign described was to maximise the UK uptake of the flu 'vaccine' which became prioritized upon Whitty's advice and in the middle of the supposed 'COVID-19 Pandemic', at a time when the ONS reported that flu deaths and had virtually flat-lined.

Matt Hancock

The Bernician similarly performs an adept analysis of the role of Matt Hancock, UK Secretary of State for Health & Social Care [DHSC], in the COVID fraud.[283]

As head of DHSC Hancock was in charge of prevention and control of 'pandemics' and maximising 'vaccination' uptake. On January 24, 2019, Matt Hancock met Bill Gates at The World Economic Forum to discuss infection control at the global level.

The details of the discussion have been subject to an FOI request, but the Cabinet Office replied that the information requested was not held on record.[284]

One year after their first meeting, on 24/01/2020, the 1st defendant held a ministerial meeting at the DHSC with Bill and Melinda Gates, to discuss antimicrobial resistance and research.[285]

A FOI request revealed that the DHSC holds no information as to the details of the discussion between Hancock and Gates, so it is reasonable to infer they discussed the pandemic and a vaccination plan. Between those two meetings with Gates, ministerial records show that the 1st defendant held numerous ministerial meetings with GSK, the Wellcome Trust and other representatives of the

[283] https://www.thebernician.net/the-defendants-their-gates-connections-matt-hancock/

[284] https://www.whatdotheyknow.com/cy/request/davos_matt_hancock_bill_gates_me

[285] https://assets.publishing.service.gov.uk/government/uploads/system/uploads/attachment_data/file/810587/dhsc-ministerial-transparency-returns-meetings-jan-mar-2019.csv/preview

stakeholders of the 'vaccine' industry.[286]

Hancock stated in a press release on April 17, 2020, in relation to the launch that day of the UK 'Vaccine' Taskforce:

> *We're doing everything possible to save lives and beat this disease, and that includes working flat out with businesses, researchers and industry to find a vaccine as quickly as possible.*
>
> *The UK is world-leading in developing vaccines. We are the biggest contributor to the global effort—and preparing to ensure we can manufacture vaccines here at home as soon as practically possible.*[287]

Then, on July 13, 2020, the Independent published the following statements:

> *Matt Hancock has said the UK will see the biggest seasonal flu vaccination drive in history this winter to try and mitigate the fear of a winter crisis mixed with coronavirus.*
>
> *Speaking remotely at the National Pharmacy Association annual conference, Mr Hancock said the government was planning now for winter, adding: "We're going to frankly need to use all of the capabilities at our disposal to deliver the vaccine programmes that we need to in the months ahead."*

---

[286] https://assets.publishing.service.gov.uk/government/uploads/system/uploads/attachment_data/file/810587/dhsc-ministerial-transparency-returns-meetings-jan-mar-2019.csv/preview

[287] https://www.gov.uk/government/news/government-launches-vaccine-taskforce-to-combat-coronavirus

Hancock stated that:

> *We're working hard on a combination of the Covid vaccination programme, should a vaccine work, and of course, the science on that is as yet unproven. And of course, the biggest flu vaccination programme in history. I want to see pharmacies involved in that flu vaccine roll-out.*[288]

Hancock's words confirm that his ministerial department launched the biggest campaign in history to maximise flu 'vaccination' uptake, just four months into the 'COVID-19 Pandemic'. It also confirms that the DHSC was planning to apply the same strategies to maximise the uptake of the fast-tracked COVID-19 'vaccines', in which Hancock declared that the UK was taking a leading role.

Such evidence above shows that Bill Gates and Matt Hancock had both the motive and the opportunity to agree that the 'COVID-19 Pandemic' would be used to maximise 'vaccination' uptake in the UK and the rest of the world.

Professor Neil Ferguson

As a government 'expert' on pandemic response, Imperial College, London professor, Neil Ferguson has an abject history of failure. As far back as 2002 Ferguson was exaggerating outcomes from infectious diseases. He predicted that up to 50,000 people would die from variant Creutzfeldt-Jakob, better known as "mad cow disease", increasing to a prediction of 150,000 if the epidemic expanded to include sheep.

He was utterly wrong.

According to the National CJD Research and Surveillance Unit at the University of Edinburgh: "Since 1990, 178 people in the

[288] https://www.independent.co.uk/news/health/coronavirus-matt-hancock-flu-pharmacists-nhs-vaccination-uk-a9615981.html

United Kingdom have died from vCJD."[289]

In 2005, the scaremongering professor claimed that up to 200 million people would be killed by bird-flu or H5N1. By early 2006, the WHO had only linked 78 deaths to the 'virus', out of 147 reported cases.[290]

Thereafter, Ferguson and his team at Gates and Wellcome Trust funded Imperial College advised the government that the 2009 swine flu or H1N1 would probably kill 65,000 people in the UK. In the end, swine flu is recorded to have claimed the lives of 457 people.[291] Notwithstanding Ferguson's proven track record of either incompetence or dishonesty, he has long been the head of the WHO Collaboration Centre for Infectious Disease Modelling.

Referring to Ferguson as "the 4th Defendant" in a supposed criminal case, The Bernician tells us that the Ebola Pandemic in Sierra Leone in 2014 was a dummy run for the Covid caper. He writes:

> *In the words of the 2nd and 4th defendants, six months after the purported outbreak, the strategies they devised to deal with the Ebola 'pandemic' in Sierra Leone were distinctly reminiscent of the strategies the 1st, 2nd and 3rd defendants proposed to the UK Government, upon the advice of the 4th defendant, six months into the 'COVID-19 Pandemic'.*[292]

In fact, some might see Ebola as a dummy run for COVID-19, given

[289] https://med.uth.edu/blog/2017/01/05/new-research-could-lead-to-blood-test-to-detect-creutzfeldt-jakob-disease/

[290] https://www.prb.org/avian-flu-and-influenza-pandemics/

[291] https://www.spectator.co.uk/article/six-questions-that-neil-ferguson-should-be-asked

[292] https://www.thebernician.net/the-defendants-their-gates-connections-neil-ferguson/

that it involved three of the defendants in this case, in almost identical circumstances. It is certainly worthy of note that, just before the Ebola 'outbreak' in the spring of 2014, the Sierra Leone Government introduced the Rotavirus 'vaccine', upon the recommendation of Gates funded GAVI and the WHO.

The Rotavirus 'vaccine' was manufactured by GSK, which started developing it in 2013, one year after the 3rd defendant became the company's President of Research and Development and the same year GSK and the Gates Foundation formed a partnership which continues to this day.[293]

The evidence is also clear that Ferguson is in the pocket of Bill Gates. According to the Gates Foundation website, in March 2020, Ferguson's employer, Imperial College, received $79 million from the foundation, for the purposes of pursuing research into the development of a new tool for malaria control and elimination in Sub-Saharan Africa.

However, this comprises only the tip of an iceberg of Gates funding received by Imperial College and Wellcome Trust during the 1st defendant's tenure.[294]

The Bernician's research uncovered that Ferguson's Imperial College Model was generated under the auspices of the Vaccine Modelling Impact Consortium, hosted by Imperial College—both effectively funded by Bill Gates and Britain's Wellcome Trust.

The VIMC is hosted by the Department of Infectious Disease Epidemiology at Imperial College. It is funded by the Gates Foundation and by "GAVI, the vaccine alliance" (GAVI's own title for itself).

Gates began funding Imperial College in 2006, four years before the Gates Foundation launched the Global Health Leaders

---

293 https://path.azureedge.net/media/documents/VAD_rotavirus_sierra_leone_fs.pdf

294 https://www.gatesfoundation.org/How-We-Work/Quick-Links/Grants-Database#q/k=Imperial%20College

Launch Decade of Vaccines Collaboration (GHLLDVC) and four years after the 4th defendant had demonstrated his penchant for overblown projections on mortality numbers from H5N1.

Up to the end of 2018, the Gates Foundation had sponsored Imperial College with $185 million. That makes Gates the second largest sponsor, second only to the Wellcome Trust, a British 'vaccine' research charity, which, by the end of 2018, had already provided Imperial with over $400 million in funding.

In December 2018, CEPI went into partnership with Imperial College. CEPI provided funding of $8.4 million for Imperial to work on a 'vaccine' platform that can be used to "rapidly develop vaccines against pathogens—even unknown ones."

An Imperial College statement claimed that the partnership of CEPI and IC aimed to develop 'vaccines' "against new and unknown pathogens within 16 weeks from identification of antigen to product release for trials". This is an extraordinary claim, when 'vaccines' previously had a typical R&D gestation period of up to fifteen years before being safely approved for public consumption.

Chapter 37—Vaccine Passports: A Chinese-Style Social Credit System

WE ARE TOO willing to accept the narrative that compels us to surrender our freedoms to government and have them given back piecemeal if we surrender to the vaccine passports.

Let us be clear. We will never be free if we accept this. For reference, simply look at China where the nation has a tight Social Credit System, a national credit rating and blacklisting run by the communist government of the People's Republic of China. This program was first initiated in regional trials in 2009. China's social credit scheme is developing, but it is only one part of the country's surveillance state. A vast network of 200 million CCTV cameras across China ensures there's no dark corner in which to hide.

Currently, the technology is reliant on Smartphone apps used to collect data and monitor online behavior on a day-to-day basis. But, as scientists are showing, the COVID jabs appear to contain graphene particles which may be the transhumanist next step to placing such apps inside our bodies.

China has been the testing ground for this dystopian future and the most compliant among us get top "citizen scores" allowing them VIP treatment at hotels and airports, cheap loans and a fast track to the best universities and jobs. But for those further down the ladder, those who do not comply with government diktat, expect to be locked out of society and banned from travel, or barred from getting credit or government jobs.

As the World Council for Health (WCH) tells us:

> *The right of bodily integrity and the right to informed consent are inalienable and universal human rights, which have been trampled by*

> *government mandates and corporate imperatives. Thus, the World Council for Health declares that any person or organization directly or indirectly participating in the manufacturing, distribution, administration or promotion of Covid-19 experimental biologics will be held liable for the violation of principles of justice grounded in civil, criminal, constitutional and natural law, as well as international treaties.*[295]

What is now apparent to those who have worked hard to determine what is happening, and who are not under the spell of the gaslight, is that the COVID-19 scam includes an extremely well-financed, carefully planned, highly organized effort to wreck society as we know it.

The COVID-19 *pandemic* was declared in March of 2020 by the corrupt corporate lobby known as the World Health Organization (WHO). We know of their corruption from years ago. In 1959, WHO entered into a secret agreement with the nuclear lobby IAEA so that the nuclear industry may control what the public heard from WHO about radioactivity's impact on human health.[296]

Fast forward to March 2020 when WHO declared a pandemic, even though there was no pandemic in any normal sense of the word. We know this because at the time supposed cases were 44,279 and of that resulted 1,440 deaths supposedly due to COVID, outside of China. The details were recorded by Professor Michel Chossudovsky in his COVID timeline at globalresearch.ca.

Thereupon, the mainstream media went into overdrive to promote the falsehood of a pandemic and the lie was repeated daily in every prominent media outlet.

---

[295] https://worldcouncilforhealth.org/campaign/covid-19-vaccine-cease-and-desist/#full

[296] https://ratical.org/radiation/radioactivity/ToxicWHO+IAEA.html

What was not being reported was that that very definition of a pandemic had been altered in 2009 by WHO. In effect, from that year onwards, the word pandemic may now be ascribed to any minor global disease. The term pandemic was thereby the trigger word for mass governmental implementation of junk science policies because each state could now justify a state of emergency because the term pandemic was being officially applied by that corrupt body, WHO.

Almost immediately, political and medical authorities brought about new measures to limit our freedoms that were, in effect, akin to medical martial law. That most citizens obediently complied in a wave of civil unity to *flatten the curve* and avoid overwhelming hospitals was a genius stroke to use our own good nature against us.

Small companies closed, but big corporate concerns remained open, as if the virus chose not to access big businesses and only plagued smaller ones.

Out the window went Constitutions and Common Law, and basic human rights. Insanely, religious and political gatherings became dangerous to health, but frequenting liquor stores was not. Locking down people from visiting family and loved ones wreaked havoc on mental health and the true pandemic was the increased stress, depression, drug abuse, and family financial problems. But ironically, with everyone in lockdown many took the time to investigate what was really going on.

Those with attuned critical reasoning skills became aware that the word case rather than the word death became the metric for determining the severity of the situation and to rational minds, that raised suspicions.

Hospitals were no longer reporting to cases of pneumonia, influenza—everything became COVID!

As Robert Snefjella wrote in Global Research:

> *...in the spirit of the gaslight, the word case has been used in two nefarious ways: First, as an outright false identification of the nebulous*

> *COVID-19; second, as the dubious result of anecdote, or of the extremely dubious result of testing, often a PCR test.*
>
> *The conjuring up of requisite numbers of so-called cases has been used to 'justify' lockdowns, quarantines, tracking, surveillance, economic turmoil, and general hysteria at home and school and work and play and travel and at church and so on, in many countries.*
>
> *While authorities cited so-called cases which were not cases in normal usage of the word, the public simultaneously was inundated with the falsehood that there were no effective treatments for actual cases of illness that had symptoms of a new flu.*
>
> *But as Sen. Dr Rand Paul (R-KY) has pointed out numerous times on the harm Dr Fauci and other "experts" have done by not promoting early treatment for Covid. "I would venture to say that thousands of people die in our country every month now from Covid because [Fauci] deemphasized the idea that there are therapeutics.*[297]

The mendacious referencing of cases and not actual deaths at certain times during the fake pandemic has been noted by many. For instance, In April 2020, the media focused on % death rates. In January 2021, the media focused on deaths per day. In December 2021, the media focused on cases.

We may thus fairly infer that for over 20 months, the media focus was on any convenient statistic that will scare people the most. Because fear sells.

---

[297] https://www.globalresearch.ca/genocide-gaslight-predatory-finance-ultimate-atrocity/5764125

By 2021, as the rollout of the COVID-19 vaccines ran apace critical thinkers became increasingly irked by the constant use of the term 'vaccine.' This was no such thing by any traditional meaning of the term. Indeed, the 'vaccines' offered no real lasting immunity beyond a few months, they did not prevent infection or transmission and in authorities were forced to clutch as straws in making the admission that, at best, the jabs may be only lessening the severity of some cases.

Instead of focusing on deaths, all the politicians and media did relentlessly and mendaciously was to bang the drum about cases, cases, cases.

Robert Snefjella, in his excellent piece for Global Research, wrote:

> *The word case in normal usage in medicine signifies some actual illness or disease: Reference to cases of pneumonia, or cases of mumps, means people actually had pneumonia or mumps. However, in the spirit of the gaslight, the word case has been used in two nefarious ways: First, as an outright false identification of the nebulous COVID-19; second, as the dubious result of anecdote, or of the extremely dubious result of testing, often a PCR test.*
>
> *The conjuring up of requisite numbers of so-called cases has been used to 'justify' lockdowns, quarantines, tracking, surveillance, economic turmoil, and general hysteria at home and school and work and play and travel and at church and so on, in many countries.*
>
> *While authorities cited so-called cases which were not cases in normal usage of the word, the public simultaneously was inundated with the falsehood that there were no effective treatments for actual cases of illness that had symptoms of a*

new flu.[298]

As for the contents of these secret potions, Dr Luis De Benito found Bluetooth connectivity in a large majority of jabbed people, and none in the un-jabbed.[299]

Reports abounded that many thousands of people were finding that magnets and ferrous metal would stick to their bodies after having been given the bio-weapon.[300]

In Japan, Moderna vials were found to contain contaminants.[301]

Graphene oxide has been found by various researchers.[302]

Other researchers have found parasites.[303]

In effect, these jabs were over-riding natural immunity rather than enhancing or being compatible with it.

To some researchers, this is proof that the jab was a eugenics tool to kill or severely impair and shorten the life of the jabbed.

> *Those who can make you believe absurdities, can make you commit atrocities.*
> —**Voltaire**

---

[298] https://www.globalresearch.ca/genocide-gaslight-predatory-finance-ultimate-atrocity/5764125

[299] https://www.bitchute.com/video/s9k346hZG0OD/

[300] https://www.bitchute.com/video/9gGUIRrlfyeo/

[301] https://www.naturalnews.com/2021-08-30-contamination-japan-discovers-black-substances-moderna-vaccines.html

[302] https://dailyexpose.uk/2021/08/30/american-scientists-confirm-toxic-graphene-oxide-and-more-in-covid-injections/

[303] https://www.bitchute.com/video/VrOjAcV5yesH/

Chapter 38—Legal Remedies Pursued: Kennedy

IT WOULD TAKE tens of thousands of words and hundreds of additional pages to give a full account of all the brilliant and brave lawyers pursuing justice against the COVID-19 tyranny.

We will focus on just those few, Dr Reiner Fuellmich in Germany, Robert Kennedy Jr. Kennedy in the U.S. and Lois Bayliss in England, to whom Principia Scientific International (PSI) has been especially supportive.

Robert Kennedy Jr.

Prominent in the legal war against vaccine tyranny is Robert Francis Kennedy, Jr. Kennedy is an American environmental lawyer and son of U.S. Senator Robert F. Kennedy and a nephew of President John F. Kennedy. He has been successful in litigating to expose how corrupt government protects Big Pharma vaccine interests.

Kennedy gave an insightful Youtube interview to Mikhaila Peterson in which he set out, from a legal perspective, the actual risks posed by the glut of untested vaccines from the pharmaceutical companies.[304]

Below is an excerpt from the interview, titled: *Opposing Views: Are Vaccines Safe?* written by Robert F. Kennedy Jr. & Amesh Adalja.

Kennedy: *Under the Vaccine Act you cannot sue a vaccine company for a vaccine injury no matter how reckless they were, how negligent they were, no matter how grievous was your injury. The only exception to that is if they knew of a vaccine injury and failed to list it.*

---

[304] https://www.youtube.com/watch?v=lkKOt4SYYiY

Peterson: *What are your views on the safety of vaccines in general?*

Kennedy: *Each vaccine has different characteristics. You have to measure each one on the safety profile of the vaccine and the risk of the infectious disease. The problem with most vaccines currently mandated for children under the CDC schedule is that none of them have been properly safety tested. Vaccines are exempt from the clinical trials of the double blind placebo—the kind of testing we apply to every other kind of medical product. So we do not know the risks of any of the vaccines, currently mandated for our children.*

I actually sued the HHS [U.S. Department of Health & Human Services] *in 2018 to show the pre-clinical safety tests for vaccines on children and after a year and a half of litigation the HHS admitted that it had never safety tested any of the vaccines. Because of that no one can say with any scientific certainty that those products are averting deaths and injuries, or causing more deaths than they are preventing.*

Peterson: *Are there some vaccines that you have seen that are more risky than others?*

Kennedy: *I would say that the most injurious vaccines are the Hepatitis B vaccines. One of the methods we can use to measure this is the Vaccine Adverse Event Reporting System (VAERS), which a post-licensure surveillance system. Unfortunately, that system is also not reliable. The HHS's own Lazarus Study found that the VAERS system captures fewer than one percent of vaccine injuries.*[305]

This proves the HHS has no idea how many people are being killed

---

305 https://digital.ahrq.gov/sites/default/files/docs/publication/r18hs017045-lazarus-final-report-2011.pdf

or injured by the vaccines. Kennedy spoke of the numerous cases that have going through the vaccine injury courts, where those who have been harmed by the jabs have sued the government and won.

Peterson: *I heard your Instagram account was shut down. Why do you think you are being targeted?*

Kennedy: *Instagram said I was passing vaccine misinformation, but they were not able to identify a single post I made that was factually erroneous. We have a huge fact checking operation at the Children's Health Defence including 312 PhD scientists.*[306]

Kennedy explained that Instagram uses the term *vaccination disinformation* as a term to cover any statement that challenges official orthodoxy. He went further to assert that anyone who challenges any pharmaceutical products is liable to be shut down and censored.

Asked by Peterson as to why this should be accepted as the norm Kennedy explained it was about profits by a coalition of Big Pharma with media giants making billions from the lockdowns.

Kennedy explained that the coalition behind the lockdowns engineered a transfer of wealth in the sum of $3.8 trillion from the middle classes globally to a handful of super rich. Kennedy named Bill Gates, Google, Facebook and Amazon among the main benefactors.

Kennedy: *They shut down around a million businesses. Who is going to profit from that? If you lock people in their homes for a year, who is going to profit?*

Ask the question: qui bono?

Kennedy explained that these companies have been very open about it, "The same companies profiting from the lockdown are

---

[306] https://childrenshealthdefense.org/

censoring critics of the lockdown."

Kennedy likened it to a coup d'état against democracy. He said we have seen the Constitution obliterated and the end of freedom of speech, the shutting down of religious practice while keeping the liquor stores open. He lamented, "The churches are in the Constitution, not the liquor stores."

Also, pertinent to the situation in the United States are those legal precedents addressing pandemics. In particular, during the 1902 smallpox epidemic, the U.S. Supreme Court in *Jacobson v. Massachusetts,* 197 U.S. 11 (1905) ruled that the State of Massachusetts could compel residents to obtain free vaccination or revaccination against the infection, or suffer a penalty of $5 (about $150 today) for noncompliance. But as professors Harvey Risch and Gerard Bradley wrote in their excellent article:

> *In authoring the majority opinion in Jacobson, Justice John Marshall Harlan argued (1) that individual liberty does not allow people to act regardless of harm that could be caused to others; (2) that the vaccination mandate was not shown to be arbitrary or oppressive; (3) that vaccination was reasonably required for public safety; and (4) that the defendant's view that the smallpox vaccine was not safe or effective constituted a tiny minority medical opinion.*[307]

[307] https://www.theepochtimes.com/covid-19-vaccine-mandates-fail-the-jacobson-test_4185648.html?utm_source=opinionnoe&utm_campaign=opinion-2022-01-09&utm_medium=email&est=vVtAFYjqFqOaQKL1n8KMqaJ45hiDbAL3%2FfSd%2BgjBIz3cKwiLQUnVgDe7%2B8%2BuGlFxH07ePeKKqKcS

Exposing the Evil Dr Anthony Fauci

In his new book Robert Kennedy Jr., *The Real Anthony Fauci* reveals all we need to know about Dr Anthony Fauci. Critically, Fauci and six co-conspirators at NIAID own patents in the Moderna vaccine, a damning conflict of interest for a government official paid by taxpayers.[308]

Thanks to the Bayh-Dole Act, government workers are allowed to file patents on any research they do using taxpayer funding.[309] Anthony Fauci, director of the National Institute of Allergy and Infectious Diseases (NIAID) since 1984. Fauci co-owns over 1,000 patents, including patents being used on the Moderna vaccine…which he approved government funding for.[310]

In fact, the NIH (which NIAID is part of) claims joint ownership of Moderna's vaccine.[311] Then we know Fauci is under intense scrutiny for illegal gain-of-function research.

What is "Gain-of-Function" research? [312] It's where scientists attempt to make viruses *gain* functions—i.e. make them more transmissible and deadlier. The U.S. government rightly banned the practice.[313]

So what did the Fauci-led NIAID do? They pivoted and outsourced the gain-of-function research (in coronaviruses no less)

---

[308] *The Real Anthony Fauci: Bill Gates, Big Pharma, and the Global War on Democracy and Public Health,* Children's Health Defense

[309] https://research.wisc.edu/bayhdole/

[310] https://childrenshealthdefense.org/defender/truth-rfk-jr-naomi-wolf-constitutional-rights/

[311] https://www.axios.com/2020/06/25/moderna-nih-coronavirus-vaccine-ownership-agreements

[312] https://ahrp.org/what-is-gain-of-function-research-who-is-at-high-risk/

[313] https://www.science.org/content/article/us-halts-funding-new-risky-virus-studies-calls-voluntary-moratorium

to China—to the tune of a $600K grant.[314]

Fauci used precisely the strategy that he developed and tested during the HIV days. Kennedy's book shows that the government's dysfunctional COVID-19 response was orchestrated by Fauci and the USG/HHS (US Government Health and Human Services Department). It is a prime example of regulatory capture and coercion of colleagues to follow a Noble Lie.

Fauci stifled open debate to the point of utter stagnation of biomedical science. Kennedy shows that the personal opinion and bias of Dr Fauci was repeatedly substituted for evidence-based medicine, and today we all live with the consequences.

---

[314] https://www.dailysignal.com/2021/04/06/fauci-must-explain-why-oversight-bypassed-for-funding-to-wuhan-lab-congressman-says/

Chapter 39—German Legal Arguments: Dr Reiner Fuellmich

IN GERMANY, LAWYERS and judges have set out legal arguments against the pandemic fraud on the web site, *Netzwerk Kritische Richter und Staatsanwälte n.e. V.* KRiStA.[315] It outlines 10 compelling reasons against compulsory vaccination against the SARS-CoV-2 virus.

10 Reasons to Oppose Compulsory Vaccination

1. Offers only self-protection

   The COVID-19 vaccination does not protect against infection and transmission of the SARS-CoV-2 virus, according to the official information from the regulatory authority EMA. All COVID-19 vaccines have been approved by the EMA only to protect against COVID-19 disease, i.e. to protect against a severe course after infection with SARS-CoV-2.

2. No correlation between infections and vaccination rates

   According to a comprehensive Harvard study, there is no correlation between infection figures and vaccination rates. On the contrary, the study even found a slight tendency that as vaccination rates increase, so do infection rates. The results of the study are in line with the negative experiences of some countries with particularly high vaccination rates (Gibraltar (about 100 %), Iceland, Ireland, Portugal), which have seen an increase in infection numbers despite high vaccination rates.

---

[315] https://netzwerkkrista.de/2021/12/10/10-gruende-gegen-die-impfpflicht/

According to this Harvard study, a positive effect of the vaccination rate on the incidence of infection cannot be proven.

3. Not comparable to measles or small pox

   The COVID-19 vaccination is in no way comparable to the measles or smallpox vaccination, because the COVID-19 vaccination precisely does not protect against infection and transmission of the virus and does not lead to sterile immunity. In contrast to measles and smallpox vaccination, a positive effect of COVID-19 vaccination on the incidence of infection with SARS-CoV-2 cannot be proven.

   Moreover, the lethality of smallpox is around 30%, the infection mortality of SARS-CoV-2 is 0.23 % on average according to the WHO. Due to the different danger, the smallpox or measles vaccination cannot be used as a comparison.

4. Vaccinated have high viral load

   According to the CDC, the English PHE health authority and four studies, the vaccinated have a comparably high viral load as the unvaccinated when they become infected. This means that vaccinated people are just as contagious as unvaccinated people, vaccinated people pose a comparably high risk of infection as unvaccinated people.

5. Many hospitalized are fully vaccinated

   According to the Robert Koch Institute (RKI), Nov. 25, 2021, 56% of hospitalized COVID-19 patients over 60 years of age were dually vaccinated. Public Health Wales reported as of 9/11/2021 that 83.6% of hospitalized COVID-19 patients were dually vaccinated. The claim that mainly unvaccinated patients are in hospital because of COVID-19 is not true.

   In the UK, by December 2021, a subtle language

change was needed in the official narrative. Because so many hospitalised patients are vaccinated, Prime Minister, Boris Johnson, has stopped referring to the 'unvaccinated' and now refers to them as people who 'haven't had their boosters.'

6. No overloading of the healthcare system

According to the report of the Federal Audit Office, there was no overloading of the health system in Germany in the first pandemic year 2020. On the contrary, there were even more hospital beds occupied in 2019 than in 2020. An analysis of the performance of hospitals of April 30, 2021 by the Advisory Council of the Ministry of Health also comes to the conclusion that on average four percent of all intensive care beds were occupied with Corona patients and that the pandemic did not bring inpatient care to its limits at any time.

According to the report of the Federal Audit Office and the analysis of the Advisory Council of the Ministry of Health of April 30, 2021, there was no overloading of the health system during the first, second and third *pandemic waves.* The question arises as to why there should now be an overload in the fourth *wave*, especially since 70% of people are now vaccinated and should therefore be protected against a severe course. Therefore, there should be no overloading of the health system at this point in time if the COVID-19 vaccinations actually offered protection against a severe illness.

7. Fewer beds occupied today than in April 2021

Even in the current situation, there is no threat of an overload of our health care system, since according to the DIVI intensive care register for which the Robert Koch Institute (RKI) is responsible, there is no increase in the total occupancy of intensive care beds. On the contrary, there are currently even slightly fewer intensive care beds occupied overall than in April 2021. Furthermore, according to the weekly report of the influenza working group, there is also no increase in acute respiratory diseases. The incidence of acute respiratory illnesses, which also includes COVID-19, is in the range of the previous years: 2017, 2018 and 2019.

8. Mortality rate like the common flu

According to the WHO epidemiological bulletin of October 2020, the infection mortality of SARS-CoV-2 is 0.23% on average. The infection mortality of 0.23% corresponds to that of influenza. In previous years, no compulsory vaccination was considered necessary during influenza waves, so the question arises as to why compulsory vaccination should be necessary now because of COVID-19. The fact that SARS-CoV-2 is less deadly than initially assumed is also confirmed by the fact that, according to Prof. Kauermann from the Institute for Statistics at the Ludwig-Maximilians University in Munich and a study by the University of Duisburg-Essen: there was no excess mortality in Germany in 2020.

9. Compulsory vaccination contradicts a free democracy

Protection of the general public through COVID-19 vaccination cannot be proven (see explanations under points 1 to 3). If compulsory vaccination were introduced only for individual protection, it would also be logical to ban high-risk sports, motorcycling, smoking, alcohol and particularly sugary drinks. Anyone who then needs medical treatment because of their high-risk lifestyle or unhealthy lifestyle would have to be denied it. This contradicts a free democratic basic order.

10. Alternative treatments exist

Compulsory vaccination would only be constitutional if there were no alternative treatment options for COVID-19, but only COVID-19 vaccination was available as a preventive protective measure. This seems doubtful since there are scientific publications according to which treatment with Ivermectin can reduce hospitalization by 75 to 85%.

Likewise, there are positive empirical values with the use of Ivermectin for COVID-19 from some Indian states, from Mexico and Peru. The testing of whether Ivermectin is suitable for the treatment of COVID-19 has not yet been completed and must not be hindered or suppressed, but this seems to be happening at present for purely financial reasons. The Bavarian State Parliament also addressed the use of Ivermectin for the treatment of COVID-19 in its resolution of June 24, 2021. As long as it cannot be ruled out that Ivermectin is an alternative treatment option for COVID-19, this argues against compulsory vaccination.

The legislator must prove that no alternative treatment options are available besides vaccination. In case of doubt, this is at the expense of compulsory vaccination.

During an interview with Dr Reiner Fuellmich, Holocaust survivor Vera Sharav draws on her experience during Nazi Germany to form her perspective on what is happening in the world today. During the interview she goes on to say:

Under the Nazi Regime, moral norms were systematically obliterated. The medical profession and institutions were radically transformed, academic science, the military, industry and clinical medicine were tightly interwoven, as they are NOW. The Nazi system destroyed a social conscience in the name of Public Health. Violations against individuals and classes of human beings were institutionalised. Eugenics driven public health policies replaced the Physician's focus on the good of the individual. [The] *German medical profession and institutions were perverted. Coercive public health policies violated individual civil and human rights. Criminal methods were used to enforce policy. Nazi Propaganda used fear of infectious epidemics to demonise Jews as spreaders of disease, as a menace to public health...Fear and propaganda were the psychological weapons the Nazis used to impose a genocidal regime and today, some are beginning to understand why the German people didn't rise up, fear kept them from doing the right thing. Medical mandates are a major step backwards towards a fascist dictatorship and genocide. Government dictates, medical intervention, these undermine our dignity as well as our FREEDOM...The stark lesson of the Holocaust is that whenever doctors join forces with government and deviate from their personal, professional, clinical commitment to do no harm to the individual, medicine can then be perverted from a healing, humanitarian profession to a murderous apparatus...What sets the Holocaust apart from all other mass genocides is the pivotal role played by the medical establishment, the entire medical establishment. Every step of the murderous process was endorsed by the academic, professional medical establishment. Medical doctors and prestigious medical societies and institutions lent the veneer of legitimacy to infanticide, mass murder of civilians. T4 was the first industrialised medical murder project in history. The first victims were disabled German infants and children under*

3...The next victims were the mentally ill, followed by the elderly in nursing homes. The murderous operations were methodical, and followed protocol very, very carefully.[316]

[316] https://www.fuellmich.com/kontakt/

Chapter 40—British Legal Arguments: Lois Bayliss

ON DECEMBER 6TH, 2021, British solicitor, Lois Bayliss and her team filed an official complaint with the Prosecutor of the International Criminal Court (ICC) on behalf of the people. Their case alleges crimes committed by UK government officials and international world leaders of various violations of the Nuremberg Code, crimes against humanity, war crimes and crimes of aggression perpetrated against the peoples of the UK.

Bayliss and her legal team tried to raise this case through the local English police and the English Court system without success and, in the final resort, have no option but to seek a legal remedy via International Criminal Court (ICC)

The complaint states:

> *The seriousness and extent of the crimes committed in the United Kingdom, highlighted by the scope of people that these crimes affect, that these crimes continue to be committed, the wide range of perpetrators, the recurring patterns of criminality and the limited prospects for accountability at the national level, all weigh heavily in favour of an investigation.*

Subject of complaint:

- Violations of the Nuremberg Code
- Violation of Article 6 of the Rome Statute
- Violation of Article 7 of the Rome Statute
- Violation of Article 8 of the Rome Statute
- Violation of Article 8 bis3 of the Rome Statute

Listed among the applicants filing the complaint is Astrophysicist activist and science adviser to Principia Scientific International for over a decade, Piers Corbyn (brother of former Labour Party Leader, Jeremy).

Listed among the *Perpetrators* of the alleged offenses are UK Prime Minister Boris Johnson, Chief Medical Officer for England and Chief Medical Adviser to the UK Government, Christopher Whitty, (former) Secretary of State for Health and Social Care, Matthew Hancock, (current) Secretary of State for Health and Social Care, Sajid Javid, Chief Executive of Medicines and Healthcare products Regulatory Agency (MHRA) June Raine, Director-General of the World Health Organisation Tedros Adanhom Ghebreyesus, Co-chair of the Bill and Melinda Gates Foundation, Bill and Melinda Gates, Chairman and Chief executive officer of Pfizer Albert Bourla Chief Executive Officer of AstraZeneca Stephane Bancel, Chief Executive Officer of Moderna Pascal Soriot, Chief Executive of Johnson and Johnson Alex Gorsky, President of the Rockefeller Foundation Dr Rajiv Shah, Director of the National Institute of Allergy and Infectious Disease (NIAID) Dr Anthony Fauci, Founder and Executive Chairman of the World Economic Forum Klaus Schwab and President of EcoHealth Alliance Dr Peter Daszack.

Setting out the background of the case the writ states:

> *The Corona virus vaccines are an innovative medical treatment, which have only received temporary Authorisation under Regulation 174 of the Human Medicine Regulations Act (2012). The long-term effects and safety of the treatment in recipients are unknown. It is important to note that the Corona Virus 'vaccines' are the world's first introduction to the synthetic m-RNA technology and all previous immunisations worked in a totally different manner, by way of introducing a deactivated or weakened virus to the*

> *body to trigger a natural arousal of the immune system against it.*

Thereafter, the case is set out that the 'vaccines' are all still within the Phase 3 COVID-19 trials which do not conclude until late 2022/early 2023.

> *The vaccines are, therefore, currently experimental with only limited short-term and no long-term adult safety data available.* [Ibid.]

Crucially, these 'vaccines' do not meet the requirements to be categorised as vaccines and are in fact gene therapy (see Appendix 8). "*The Merriam-Webster dictionary quietly changed the definition of the term 'vaccine' to include components of the COVID-19 m-RNA injection.*" [Ibid.]

The definition of vaccine was specifically changed due to the COVID-19 injection on February 05, 2021.

Submitting his testimony to the court, Dr Mike Yeadon (joint applicant) asserts that claims calling the COVID-19 injections a vaccine "*is public manipulation and misrepresentation of clinical treatment. It's not a vaccination. It's not prohibiting infection. It's not a prohibiting transmission device. It's a means by which your body is conscripted to make the toxin that then allegedly your body somehow gets used to dealing with it, but unlike a vaccine, which is to trigger the immune response, this is to trigger the creation of the toxin.*"[Ibid.]

Thereafter, the writ states that the PCR tests are wholly unreliable, and a fatal error was made requiring its use as a diagnostic tool:

> *A review from the University of Oxford's Centre for Evidence-Based Medicine (Appendix 2) found that the standard PRC test is so sensitive, that it can detect old infections by picking up fragments*

> *of dead viral cells.* [Ibid.]

A key ingredient of the '*vaccines*' is graphene oxide which is cited as a major problem as per the evidence from November 23, 2021, by Dr Andreas Noack that "*due to the nano size of the graphene oxide structures they would not show up on an autopsy as toxicologists can't imagine that there are structures that can cut up blood vessels causing people to bleed to death on the inside so they would not be looking for them, given their atomic size.*"[Ibid.]

The writ then explains that on November 18, 2020, Dr Andreas Noack was on a livestream on YouTube discussing the dangers of the COVID-19 'vaccines' when he was arrested on camera by armed German police (Appendix 41). On November 26, 2021, just hours after publishing his latest video about graphene oxide and graphene hydroxide (Appendix 42) he was attacked and murdered.

The submitted formal ICC application requests a full investigation be done into the inclusion of Graphene hydroxide in the COVID-19 'vaccines' and into the assassination of Dr Andreas Noack.

On the issue of vastly inflated death rates the legal submissions cites, "A Freedom of Information request (Appendix 43) shows us that between March and June 2020 the total number of COVID-19 related deaths in England and Wales with no pre-existing health conditions was 4,476."

However, the COVID-19 deaths publicized for the same period were recorded at a mammoth 49,607.

Bayliss and her team's submission then explains that a further way that the COVID-19 statistics have been artificially inflated is by the rebranding of the common influenza, pneumonia and other respiratory infections as COVID-19.

> *Epidemiologist Knut Wittowski, the former head of biostatistics, epidemiology and research design at Rockefeller University claims 'there may be*

> *quite a number of influenza cases included in the 'presumed Covid' category of people who have Covid symptoms (which influenza symptoms can be mistaken for), but are not tested for SARS RNA'. Those patients he argued, 'also may have some SARS RNA sitting in their nose while being infected with influenza, in which case the influenza would be 'confirmed' to be Covid.'* [Ibid.]

Data from the ONS (Appendix 45) showed that deaths in 2018 from influenza and pneumonia amounted to 29,516 and in 2019, was 26,398. However, deaths in 2020 for influenza was recorded at just 394 and pneumonia at 13,619. (Appendix 46)

Additionally, John O'Loony, a joint applicant on the request is a funeral director running his own funeral home in Milton Keynes, testified (Appendix 47) that as a funeral director he saw…

> *…a massive effort made to deliberately inflate Covid death numbers. Cancer patients and stroke victims and even one guy that was run over all ended up with Covid on their death certificate.* [Ibid.]

The application then goes on to list all the various alternative treatments identified and successfully used by independent doctors during the pandemic.

Suggesting that the pandemic was pre-planned, the application cites the Clade x and Event 201 Scenario:

> *In May, 2018, the WEF partnered with Johns Hopkins to simulate a fictitious pandemic dubbed 'Clade X' (Appendix 12) to see how prepared the world be if ever faced with a catastrophic pandemic. A little over a year later, the WEF once again teamed-up with Johns Hopkins, along with*

> *the Bill and Melinda Gates Foundation, to stage another pandemic exercise called 'Event 201' in October, 2019 (Appendix 13). Both simulations concluded that the world wasn't prepared for a global pandemic. A few short months following the conclusion of Event 201, which specifically simulated a coronavirus outbreak, the World Health Organization (WHO) officially declared that the coronavirus had reached pandemic status on March 11, 2020.*

During the actual pandemic, every scenario set out in the Clade X and Event 201 simulations came into play, including:

- Governments implementing lockdowns worldwide
- The collapse of many industries
- Growing mistrust between governments and citizens
- A greater adoption of biometric surveillance technologies
- Social media censorship in the name of combating misinformation
- The desire to flood communication channels with "authoritative" sources
- A global lack of personal protective equipment
- The breakdown of international supply chains
- Mass unemployment
- Rioting in the streets

After the nightmare scenarios had fully materialized by mid-2020, the WEF founder declared "now is the time for a great reset" in June 2021, which the writ declares rises beyond the level of mere coincidence.

More so since, according to the World Economic Forum's Klaus Schwab...

> *The pandemic represents a rare but narrow window of opportunity to reflect, reimagine, and reset our world to create a healthier, more equitable, and more prosperous future.*

The Bayliss document concludes with a strong request:

> *WE WANT TO REPEAT: It is of the utmost urgency that ICC takes immediate action, taking all of this into account, to stop the rollout of covid vaccinations, introduction of unlawful vaccination passports and all other types of illegal warfare mentioned herein currently being waged against the people of the United Kingdom by way of an IMMEDIATE court injunction.*[317]

[317] https://dailyexpose.uk/2021/12/10/uk-team-file-complaint-of-crimes-against-humanity-with-the-international-criminal-court/

Chapter 41—Shining a Disinfecting Light into the Darkness

The very word secrecy is repugnant in a free and open society.
—President John F. Kennedy

AS WE EXPLAINED elsewhere, there are many similarities between the pandemic fraud and climate fraud and secrecy over the science is a cornerstone of these criminal conspiracies. We are grateful that a key element of the common law system of English-speaking nations is the duty to make full disclosure of key evidence during litigation to allow the opposing parties to fairly ascertain the truthfulness of statements made by witnesses. This is especially important in cases of scientific fraud, where the devil is very much in the details.

PSI, while primarily an international publisher of scientific content, has not shied away from litigating on behalf of our members.

The key difference between the climate and pandemic fraud is that the former was a trial run for the latter. In the earlier instance, the vast number of organisational skills, funding, coercion, blackmail, bribery, media collusion, exploitation of ignorance, group think and civil good will was played out slowly over three decades via the UN's Intergovernmental Panel on Climate Change (IPCC), numerous international NGO's, corporate co-conspirators a compliant academia and scientifically ill-educated populace was more a trial run.

So dumbed down are our younger generations, thanks to indoctrination by academics who invariably espouse anti-libertarian, pro-government sentiments. For over forty years, coalitions of academics, governments, corporations, and world governance bodies have colluded to build a global ed-tech schooling system meant to shackle children to the transhumanist Fourth Industrial Revolution.

American Federation of Teachers (AFT) is a prime example of academic sell outs to the Rockefellers, Trilateralists, and Big Tech. The country's 2nd largest teachers' union heavily lobbied the CDC to not reopen schools. While they framed their efforts as health and safety focused, there is more than meets the eye to the union's push for indefinite remote learning.

Emails obtained under FOI rules show that the bureaucracy of this teachers' union was lobbying the CDC to roll back its school reopening guidelines. The emails show the AFT has been attempting to sway the CDC to ratchet up COVID-19 restrictions that perpetuate public education's reliance on Big Tech company privatization.

As educators pushed politicians to implement more remote learning rather than have children physically attend schools the evidence showed the nonsense of the strategy. Mainstream medical journals, such as the Lancet, have published data which finds that:

> [c]*hildren and young people remain at low risk of COVID-19 mortality.*[318]

What is the point of isolating children when there is mounting evidence that in-person schoolhouse learning does not result in rising community transmission rates of COVID-19? [319] [320]

This is even more perplexing when considering the increase in

---

318 https://www.thelancet.com/journals/lanchi/article/PIIS2352-4642%2821%2900066-3/fulltext

319 https://epicedpolicy.org/wp-content/uploads/2020/12/COVID-and-Schools-Dec2020.pdf

320 https://americansforpublictrust.org/document/cdc-responsive-records/

adolescent suicides resulting from depression caused by social isolation from school lockdowns that sequester students from their classmates.[321]

Educators should know full well that there is copious data showing that face-to-face classroom instruction has far more benefits for student learning than virtual *distance learning* through a computer screen.[322]

The rapid and monolithic global response in early 2020 to the novel coronavirus is a testament to how well the same playmakers coordinating the prior scam had honed their playbook to perfection.

As we showed earlier in this book, the same billionaire elite hell-bent on eugenicist policies were in the shadows pulling the strings at the UN (either at the IPCC or WHO). These quasi-public bodies are merely private and unaccountable clubs for the ultra-rich, in the same way the Bank of England and the Federal Reserve are not public bodies but are, in fact, private banks.

One of the most galling hypocrisies is the contempt the super elite have for enshrined human rights. Despite owning and running their private UN club to serve a dystopian mission to enslave the world under vaccine tyranny, the UN is perhaps widely hailed as the champion of the Universal Declaration of Human Rights.

The UN's own website tells us:

> *The Universal Declaration of Human Rights (UDHR) is a milestone document in the history of human rights. Drafted by representatives with different legal and cultural backgrounds from all regions of the world, the Declaration was proclaimed by the United Nations General*

[321] https://chicago.suntimes.com/2021/1/27/22253057/chicago-pediatricians-remote-learning-social-isolation-suicide-depression-cps-ctu

[322] https://www.prnewswire.com/news-releases/new-survey-majority-of-teens-say-online-learning-is-worse-than-in-person-but-only-19-think-school-should-return-to-full-in-person-instruction-301131923.html

> *Assembly in Paris on 10 December 1948 (General Assembly resolution 217 A) as a common standard of achievements for all peoples and all nations. It sets out, for the first time, fundamental human rights to be universally protected and it has been translated into over 500 languages. The UDHR is widely recognized as having inspired, and paved the way for, the adoption of more than seventy human rights treaties, applied today on a permanent basis at global and regional levels (all containing references to it in their preambles).*[323]

At the time of writing, one of the few countries to recognise this inalienable right is Japan. On the nation's COVID-19 vaccination web site it is clearly stated:

Consent to Vaccination

> *Although we encourage all citizens to receive the COVID-19 vaccination, it is not compulsory or mandatory. Vaccination will be given only with the consent of the person to be vaccinated after the information provided. Please get vaccinated of your own decision, understanding both the Effectiveness In Preventing Infectious Diseases And The Risk Of Side Effects. No Vaccination Will Be Given Without Consent. Please Do Not Force Anyone In Your Workplace Or Those Who Around You To Be Vaccinated, And Do Not Discriminate Against Those Who Have Not Been Vaccinated.*[324]

---

[323] https://www.un.org/en/about-us/universal-declaration-of-human-rights

[324] https://t.co/JSbiHiXQr7

We know that the 'vaccines' are not very effective at all. The absolute risk reduction measure (ARR, vaccine effectiveness) of ALL the COVID-19 shots is <2%. The FDA and vaccine manufacturers did not disclose this to the public, ignoring the FDA's own reporting guidelines.

The touted "95% efficacy" spoken about widely is the relative risk reduction measure (i.e. vaccine efficacy) of the infection risk between the placebo and vaccinated group. Using Pfizer's clinical trial data, the infection risk was 0.761% and 0.044% for the placebo and vaccinated group, respectively.

Both groups' infection risk was <1%! The 95% vaccine 'efficacy' represents the % decrease from 0.761% to 0.044%. But the absolute risk reduction measure, e.g. how much does the shot reduce the risk of infection, is .0 0761% -0 .044% = 0.7%. So here we are, spending trillions to mass inoculate billions around the world with shots that reduce the risk of infection by <2%.

COVID-19 vaccine efficacy and effectiveness—the elephant (not) in the room.[325]

The public's right to access all the data to make informed decisions as to be jabbed or not is inviolatable. It is part of those UN-enshrined human rights that are at the core of the claims being presented to courts by thousands of international lawyers working with Dr Reiner Fuellmich.

We are living in a key moment of history that will likely permanently shape humanity's destiny.

If these lawsuits succeed, we will have restored the concept of preserving individual autonomy and self-determination. If we fail, humanity as we know it will be condemned to a transhumanist future, a two-tier world of haves and have-nots—a tiny master race of billionaires and a disposable slave class biologically engineered into compliance and servitude.

So serious a threat to humanity and public health does the

---

[325] https://www.thelancet.com/journals/lanmic/article/PIIS2666-5247(21)00069-0/fulltext

psychopathic elite represent that the scandal is being dubbed among international lawyers as Nuremberg 2.0.[326]

The use of the term *Nuremberg 2.0* is not taken lightly with the growing weight of medical opinion affirming that the experimental vaccine program is nothing short of bioterrorism.

This is the considered opinion of Dr Peter McCullough, who insists that the spike protein in COVID-19 shots is pathogenic.

Dr McCullough announced:

> *You are about five times as likely to die of the vaccine than you are to take your risks with COVID-19. Therefore, those who 'chose not to get the vaccine,' in fact 'made a smarter choice.'*

McCullough says that this lethality was not a mistake and that it is the reason for autoimmune attacks some people have experienced after accepting an mRNA vaccine.[327]

McCullough made the remark during an October 27, 2021, meeting in Phoenix, Arizona in which he explained the dangers of mRNA technology used to combat COVID-19. Reporting on the event, lifesitenews.com reported:

> *His presentation included documentation demonstrating that all age groups are more likely to die from the vaccines than from COVID-19, that the vaccinated are just as much a virus transmission threat to others as the unvaccinated, and that those who have recovered from the disease have a 56% greater chance of severe side-effects should they afterwards take the jab.*

---

[326] https://nurembergtrials.net/

[327] https://www.lifesitenews.com/news/covid-jabs-came-from-bioterrorism/

McCullough highlighted a June 2021 paper he co-authored with an international team of another 56 scientists said:

> *SARS-CoV-2 mass vaccination: Urgent questions on vaccine safety that demand answers from international health agencies, regulatory authorities, governments and vaccine developers.*
>
> *There has been no testing to see if [mRNA or adenoviral DNA] incorporates into the human genome, if it causes birth defects, or cancer.*
>
> *If we don't have the proper safety mechanisms in place, we think we're going to be in trouble.*
>
> *There have been no safety committees. [This is] extremely worrisome.*[328]

McCullough addressed COVID-19 jab mandates instituted by employers across the public and private sectors and encouraged his audience to stand firm against coercion.

Naturally, he is aware that terrible injustices have been done against those who contradict the mainstream COVID-19 narrative. Earlier in the presentation, the researcher mentioned that he had recently been stripped of two professorships which he described as hard to get.

He explained:

> *And in order to be removed as a professor, there has to be senate faculty hearings, and presentations and acts of moral turpitude, and everything else.*
>
> *Shots are research; it's 'illegal' for doctors to*

[328] https://normanpilon.com/2021/07/18/sars-cov-2-mass-vaccination-urgent-questions-on-vaccine-safety-that-demand-answers-from-international-health-agencies-regulatory-authorities-governments-and-vaccine-developers-roxana-bruno-pet/

> *encourage or discourage injections.*

McCullough explained that, because the COVID-19 vaccines are still *research,* doctors should never have encouraged their patients to—or discouraged them from—taking the vaccine.

> *If I had a research project, and saw a patient, and said, 'Listen, you should be in my research project,' do you know I would get called to the IRB?* [Institutional Review Board]
>
> *I'd probably get citations against me. It's illegal for me to do that. It violates the Office of Human Research Protections. I can't encourage people to [be part of] my research project. The COVID-19 vaccines are research.*
>
> *Nobody had a problem with the vaccines back in January because it was elective. You could do it, or you could not do it. And everybody knew it was research. The only reason we have a problem now is that it is no longer elective.*
>
> *Doctors are guilty of 'wilful misconduct' for neglecting COVID-19 patients.*[329]

McCullough told LifeSiteNews via email that data safety monitoring boards (DSMB), event adjudication committees (EAC), and Human Ethics Committees…

> *…are standard regulatory bodies agreed upon by the sponsor and regulatory authorities to oversee subject safety and integrity of the research program.*
>
> *These bodies should have been in place at the*

---

[329] https://www.lifesitenews.com/news/covid-jabs-came-from-bioterrorism/

> *very beginning of the US public program, and they should have been given a report on safety from the CDC / FDA monthly.*

Rather than wholesale rioting to put an end to this war on humanity, which would play into the hands of the dictator class, the right response is peaceably organising battalions of independent attorneys working within current legal structures to shine light where it is most needed. Our hope is to wake up the greatest number of the brainwashed and gullible.

Thankfully, most court records are public. Honest judges who uphold the law are coming back into fashion, especially since Donald Trump appointed hundreds of new judges. As westernjournal.com reports:

> *Former President Donald Trump worked hard to re-shape the federal judiciary, and now his work is paying off under the Biden administration. ...Judge Matthew Schelp of the Eastern District of Missouri issued the preliminary injunction for workers throughout 10 states at facilities that are certified for Medicare and Medicaid.*[330] *The states are Iowa, Alaska, Arkansas, Missouri, Kansas, Nebraska, New Hampshire, Wyoming, North Dakota and South Dakota.*[331]

The sanctity of our civil rights is the test for these judges who must decide if we are sovereign over our own bodies.

Stemming from a lawsuit filed in early November, Judge Kurt Engelhardt, a Trump appointee, wrote that the mandate "threatens

---

[330] https://ago.mo.gov/docs/default-source/press-releases/cms-injunction.pdf?sfvrsn=e

[331] https://www.westernjournal.com/trump-appointed-judge-smacks-bidens-covid-vax-mandate-health-workers-10-states/

to decimate [companies'] workforces (and business prospects) by forcing unwilling employees to take their shots, take their tests, or hit the road."

As he got ready to leave office, Pew Research noted:

> *Donald Trump leaves the White House having appointed more than 200 judges to the federal bench, including nearly as many powerful federal appeals court judges in four years as Barack Obama appointed in eight.*[332]

What will likely be presented to all these judges are the key human rights set out in The Nuremberg Code.

> *...a medical code of ethics based on the laws under which the Nazi criminals were judged in U.S.A. vs. Karl Brandt, et al. (Nuremberg physicians' trial), for their role in conducting horrific medical experiments during the Second World War. The Nuremberg Code later constituted the basis for the Helsinki Declaration 1965 which binds the World Medical Association and practicing physicians to 'act in the* [individual] *patient's best interest when providing medical care'.*

Informed consent to participate in a medical experiment...

The first principle of the Nuremberg Code is a willingness and informed consent by the person to receive treatment and participate in an experiment. The person is supposed to activate freedom of choice without the intervention, either through force, deceit, fraud, threat, solicitation, or any other type of binding or coercion.

Those of us with the critical reasoning skills necessary will

---

[332] https://www.pewresearch.org/fact-tank/2021/01/13/how-trump-compares-with-other-recent-presidents-in-appointing-federal-judges/

keep joining the dots to put in front of these judges. Our mission is to keep battling for transparency about the science and we are hopeful the judges will, by and large, hold the line.

Part of the evidence we will be presenting before the courts are the glaring statistical disparities which are the clue to bad faith operatives performing mathematical contortions to fool us the problem is worse than it is.

An interesting case in point is the very odd mortality results found in Scotland where there are far fewer deaths from all-causes in the 28 days following vaccination than you'd expect in the general population.

Of course, most people don't just die suddenly from COVID-19, even though that is the drama the media tries to convey. Most deaths are slow with an understanding about the inevitable, both from the individuals concerned and from their medics.

According to former CIA director William J. Casey:

> *We'll know our disinformation program is complete when everything the American public believes is false.*

Disinformation and playing fast and loose with dodgy data are what the media and politicians are very adept at. To discern the facts from the flim-flam takes a good deal of lateral thinking to unearth reliable data sources congruent to the problem.

For example, looking at Scotland's post-vaccination deaths data, as well as other data such as the UKHSA all-cause deaths by vaccination status, it becomes evident that doctors there did not vaccinate those close to death (i.e., sparing them from the potentially nasty side effects). This explains the majority of the deaths after the vaccination period (when the vaccine efficiencies were first estimated for the general population), whether 'with COVID' or from any other cause. As we write this book, we are not yet out of this effect, with a 'residual protection from hospitalisation/death' indicated for the vaccines.

What is interesting is what we'll see in the data once this effect has worn off completely—we'll see this in the unvaccinated first, as those that were close to death will have died leaving only the healthy 'vaccine hesitant' in that group.

All this wouldn't have happened if there had been implemented proper post-vaccination impact studies with a matched cohort-based approach. But that did not happen, so we are stuck wading through complex statistical effects that can occur without the population realising that there is far more going on than meets the eye.

Some might say that they did not do apply post-vaccination vigilance because it is far easier for the authorities to deny something where there's no data supporting it. That is, if vaccines actually did work they could say *we saved you!* whereas if it turns out that they made things worse they can legitimately say *but no-one knew!* This would all be very clever and certainly in the interests of the politicians, but it would certainly not be in the interests of the population in general.

Moreover, we believe there is enough data from public sources to confirm that the vaccines are not only ineffective in providing immunity, but are actually causing severe adverse reactions and deaths, notably among the young.

To that effect, data is beginning to emerge from life insurance companies of a stark rise in deaths among young people. An Indiana life insurance CEO says deaths are up 40% among people ages 18-64. The head of Indianapolis-based insurance company OneAmerica said the death rate is up a stunning 40% from pre-pandemic levels among working-age people.

The company's CEO Scott Davison said:

> *We are seeing, right now, the highest death rates we have seen in the history of this business—not just at OneAmerica. The data is consistent across every player in that business and most of the claims for deaths being filed are not classified as*

> *COVID-19 deaths. What the data is showing to us is that the deaths that are being reported as COVID deaths greatly understate the actual death losses among working-age people from the pandemic. It may not all be COVID on their death certificate, but deaths are up just huge, huge numbers.*

Davison said at the same time, the company is seeing an uptick in disability claims, saying at first it was short-term disability claims, and now the increase is in long-term disability claims.

> *For OneAmerica, we expect the costs of this are going to be well over $100 million, and this is our smallest business. So, it's having a huge impact on that.*[333]

333

https://web.archive.org/web/20220102200839/https://www.thecentersquare.com/indiana/indiana-life-insurance-ceo-says-deaths-are-up-40-among-people-ages-18-64/article_71473b12-6b1e-11ec-8641-5b2c06725e2c.html

Chapter 42—A Conspiracy for Genocide, Toxic 'Vaccine' Tyranny

SO, WHY DO we believe that there was a conspiracy among the globalist elite to inflict on a catastrophe narrative to enslave us all for their self-serving ends?

To begin your journey down that rabbit hole, a useful bitchute.com video to watch to glean some insight is *Fauci and Event 201: Clear Evidence of Foreknowledge and Plandemic Preplanning.*[334]

The above offers a snippet from a mass of evidence in documents, videos, whistle-blower testimony, leaked emails, etc. which ties into the openly stated objectives of Agenda 21, to bring about one world unelected global government that imposes severe population controls.

The common mantra, the cover for a fascistic new order is *The pairing of public and private*, which is the basic definition of fascism, and is mentioned many times as a necessity for every nation to become subsumed under a re-visioned *global community.*

Essential to bringing about the Big Lie is softening up the public before hand—grooming the masses to be willing to accept governmental diktat at times of mass panic upon a declared international emergency. We urge readers to look up Operation Mockingbird and see how media brainwashing is a known and admitted tool of mass deception by government.

In brief, Operation Mockingbird was the CIA program used to intentionally spread misinformation on American media. CIA director, William Colby, testified to the Church Committee that over 400 CIA agents were active in the U.S. media to control what

[334] https://www.bitchute.com/video/ikb6wKFsiCIy/

was reported through American mainstream media television, newspapers, radio, and magazines.

Mockingbird marked the official start of the US government's infiltration of the news and entertainment media as editors, writers, reporters, and film makers were recruited as intelligence assets, made to sign secrecy oaths, and thereafter were under the control of their new masters.

Of course, our detractors will dismiss any and all of this as mere *conspiracy theories*, but so fearful of us are the elite that any highly credentialed whistle-blowers are swiftly de-platformed.

In a 2008 Harvard law paper, *Conspiracy Theories,* Sunstein and co-author Adrian Vermeule, a Harvard law professor, ask, "What can government do about conspiracy theories?"

In that insightful 30-page paper, authors Sunstein and Vermeule, showed:

> *We can readily imagine a series of possible responses. (1) Government might ban conspiracy theorizing. (2) Government might impose some kind of tax, financial or otherwise, on those who disseminate such theories.*[335]

Sunstein allowed that:

> *...some conspiracy theories, under our definition, have turned out to be true. The Watergate hotel room used by Democratic National Committee was, in fact, bugged by Republican officials, operating at the behest of the White House. In the 1950s, the CIA did, in fact, administer LSD and related drugs under Project MKULTRA, in an effort to investigate the possibility of 'mind control'*

[335] https://papers.ssrn.com/sol3/papers.cfm?abstract_id=1084585

And, as the pandemic narrative collapses that, too, is becoming recognised as the biggest, more impactful conspiracy in human history. In a recently released book, *On Rumors*, Sunstein argued websites should be obliged to remove false rumors while libel laws should be altered to make it easier to sue for spreading such rumors.

But as Principia Scientific International proved in 2019, the libel laws worked in favor of we conspiracy theorists, when our founding chairman, Dr Tim Ball, won the science trial of the century in defeating UN climate data faker, Michael Mann's libel claim in the Supreme Court of British Columbia.[336]

A good deal of the population has woken up to the Chicken Little scaremongering about climate change. We need to ensure many more join us in the far more consequential Great Awakening about the current medical tyranny.

Exposing the COVID-19 pandemic, the fake virus, fake tests and fake vaccines is part of a great awakening, literally unfolding right before the uninformed masses how the Malthusian elite want 90% of us all dead—gone from "their" planet.

Here are some of the factual giveaways: on January 10, 2017, in a public presentation at Georgetown University, Dr Anthony Fauci made his tell-tale speech on pandemic preparedness.

He said:

> *If there's one message I want to leave with you today based on my experience is that there is no question that there will be a challenge coming to the current Administration in the arena of infectious diseases, both chronic infectious diseases…but also there will be a surprise outbreak.*

---

[336] https://principia-scientific.com/breaking-news-dr-tim-ball-defeats-michael-manns-climate-lawsuit/

Did the lead member of President Donald Trump's coronavirus task force, National Institute of Allergy and Infectious Diseases Director Dr Anthony Fauci, hold the gift of seeing into the future?

On October 29, 2019, Dr Fauci alongside Rick Bright, Director of HHS Biomedical Advanced Research & Development Authority (BARDA) spoke publicly about a virus outbreak in China spreading worldwide which would need to be tackled with innovative vaccine mass treatment.

Then on November 25, 2019, Dr Fauci "inspirationally hoped to be able to rapidly respond to something brand new, be it a pandemic or bioterror."

On October 18, 2019, came the Event 201 pandemic exercise, addressing pandemic preparedness as and the solutions the global elite intended between the public and private sectors to be the solution to the problem they created themselves befitting a classic Hegelian Dialectic scenario.

Hegelian dialectic is PROBLEM→REACTION→SOLUTION and it works like this:

- The government creates or exploits a problem in which attributes blame to others.
- The people react by asking political leaders for protection and help (safety and security) to help solve the problem.
- Then, the government offers the solution that was planned by them long before the crisis occurred.

What's the outcome? The outcome of all of this is: the rights and liberties are exchanged for the illusion of protection and help.

Event 201, coming right before the actual pandemic, was a trial run. Key speakers at Event 201 explicitly define how and why, upon declaration of such an emergency, all media and social platforms must be subsumed with one narrative for the greater good.

Key to coordinating and disseminating the "approved"

information to spin the narrative was Johns Hopkins Center for Health Security, the World Economic Forum and the Bill and Melinda Gates Foundation.

Key mRNA contributor Dr Robert Malone, a prominent skeptic of mandatory COVID-19 vaccinations, spoke to the effectiveness of the planning and preparedness of the above when he called the outcome a mass formation psychosis.[337]

Speaking on the popular Joe Rogan podcast, Malone said:

> *Our government is out of control on this and they are lawless. They completely disregard bioethics. They completely disregard the federal common rule. They have broken all the rules that I know of, that I've been trained* [in] *for years and years and years.*

Malone was de-platformed on social media for being one of the most credible key experts brave enough to risk his career to speak out.

A spokesperson for Twitter had told the left-wing Daily Dot outlet that Malone's account…

> *…was permanently suspended for repeated violations of our COVID-19 misinformation policy…per the strike system outlined here, we will permanently suspend accounts for repeated violations of this policy.*

---

[337] https://www.theepochtimes.com/mkt_morningbrief/dr-robert-malone-to-rogan-us-in-mass-formation-psychosis-over-covid_4189087.html?utm_source=morningbriefnoe&utm_medium=email&utm_campaign=mb-2022-01-03&mktids=8e27e86280e1cb6da02ffcddc523bb5b&est=TmO9H07c1w1BhWL8bmgoMrEFlaZS000ptGEYzBTIePxlluV4HdhrMOvg2Mzh6YH62SCGgKPd

Misinformation? Dr Malone is a renowned expert on mRNA vaccine technologies and received training at the University of California-Davis, UC-San Diego, and the Salk Institute.

Undeterred, Malone insisted: "These mandates…are explicitly illegal" and "are explicitly inconsistent with the Nuremberg Code. They are explicitly inconsistent with the Belmont report. They are flat-out illegal, and they don't care." Malone was referring to the 1978 report published in the Federal Register regarding ethical principles and guidelines for research involving human subjects.

Malone drew parallels to the mentality that developed among the German population in the 1920s and 1930s. In those years, Germans "had a highly intelligent, highly educated population, and they went barking mad," Malone said.

> *When you have a society that has become decoupled from each other and has free-floating anxiety in a sense that things don't make sense, we can't understand it. And then their attention gets focused by a leader or series of events on one small point, just like hypnosis. They literally become hypnotized and can be led anywhere. They will follow that person. It doesn't matter if they lie to them or whatever.* [Ibid.]

Chapter 43—Government Interventions

THE NUMEROUS MEASURES put in place by governments around the world are, by and large, not only infringements of their nation's existing laws but run counter to the very tenets set out in the Universal Declaration of Human Rights.

Such actions as long-term enforced medical isolation, restrictions on access to services of anyone unvaccinated, are not evidence-based or scientifically valid, but have significant and lasting consequences that do not contribute to improved public health. Such policies, under the guise of temporary emergency measures, are being decided using a flawed strategy.

As we showed earlier in this book, the PCR test is not a diagnostic test and therefore does not confirm illness or disease. It is of concern to many that the inventor of the test, Kary Mullis died in August 2019, not long before the pandemic took hold in China. Mullis had been a vocal critic of Dr Anthony Fauci. Mullis stated that no infection or illness can be accurately diagnosed with PCR.[338]

As such, it should not be used as a marker by which to identify those at risk. Worse yet, it is not normal for healthy people with no symptoms to constantly test themselves to see if they are sick. Governments used to lock people up in asylums for that.

As evidence of the lack of faith in the PCR test, from December 31, 2021, the CDC withdrew the request to the U.S. Food and Drug Administration (FDA) for Emergency Use Authorization (EUA) for its use in detecting COVID-19.[339]

The CDC admitting that the PCR test cannot tell the difference between ordinary influenza and COVID-19. Also, during

---

[338] https://www.globalresearch.ca/kary-b-mullis-no-infection-illness-can-accurately-diagnosed-pcr/5757224

[339] https://www.cdc.gov/csels/dls/locs/2021/07-21-2021-lab-alert-Changes_CDC_RT-PCR_SARS-CoV-2_Testing_1.html

December 2021 the battle lines were further drawn when the World Council for Health announced:

> *The World Council for Health demands an end to this crisis and hereby declares it illegal and unlawful for anyone to participate, directly or indirectly, in this harmful experimental vaccination programme. The World Council for Health declares individuals, governments, and other corporations will be held liable for their involvement.*

Building on an emerging consensus of medical dissenters, as first set out in the Great Barrington Declaration, affiliated independent doctors, scientists, researchers and legal advocates make similarly strong-worded rejoinders. For example, in the U.S., the Association of American Physicians and Surgeons (AAPS) fought mandates through Calls to Action to members and the public. AAPS physicians participated in many events and podcasts. AAPS representatives met with legislators in Washington DC specifically to discuss the vaccine mandates, OSHA requirements, federal and healthcare workers restrictions, child vaccines, etc.

Chapter 44—Adverse Drug Reactions (ADR) Monitoring; A Regulatory Scientific Trickery

THE STATEMENT FROM the Health Canada website may explain the origin of the ADR monitoring approach:

> *All marketed drugs and health products have benefits and risks. All health products are carefully evaluated before they are licensed in Canada. However, some adverse reactions or problems may become evident only after a product is in use by consumers.*[340]

So, in reality, the ADR system is an approach to monitor extremely rare and missed adverse effects during clinical trials. It is like a consumer reporting system commonly managed by a third party to assess a company, person, or object for another company or person. It is to establish the authenticity of claims of an entity for investment or purchase.

The critical point is: the ADR reporting company has to be independent of both parties to provide unbiased reporting.

The purpose of the ADR reporting party is not to make the judgment call but to report adverse effects described by consumers or, in this case, drug users (the patients). Physicians, hospitals, drug manufacturers, and regulatory authorities should not be part of the input because they are part of drug development, assessment, or sale of the products.

However, unfortunately, in the case of pharmaceuticals,

---

[340] https://www.canada.ca/en/health-canada/services/drugs-health-products/medeffect-canada/adverse-reaction-reporting.html

including vaccines, this is not the case. Here physicians and regulatory authorities who approve the drugs also become the ADR monitoring authorities and parties. Therefore, by default, ADR monitoring becomes biased and unresponsive to the need or concerns of the patients, which could be manipulated.

An example of a biased interpretation of such monitoring is reflected below (from Health Canada):

> *…each year, more drugs and natural health products are included in the adverse reactions reported to Health Canada from 25,668 reported products in 2010 to 208,383 reported products in 2019, an 8-fold increase. This may be due to improved reporting mechanisms and increased general awareness of the risks with the reporting of more drugs and natural health products …*[341]

It implies that the reported ADR may be linked to an improved reporting mechanism, not the increase in adverse effects or reports.

One should remember that in reality, the ADR monitoring system has become a regulatory setup to support the views and policies of bureaucracy, not an unbiased evaluation mechanism as intended.

For example, as described in an excellent report by Dr. Altman called *The Time of Covid*:

> *These are some of the reasons why it is widely acknowledged all adverse event reporting systems suffer from notorious underreporting. This can result in an underreporting factor of between 10-30 or more, i.e., one must multiply the official*

[341] https://www.canada.ca/en/health-canada/services/drugs-health-products/reports-publications/medeffect-canada/adverse-reactions-incidents-recalls-2019-summary.html#a2.1

> *incidence of adverse events by 10-30, to obtain a real-world estimate of the true incidence of the adverse event. For US VAERS reporting in respect of the Covid-19 'vaccines', the underreporting factor (URF) is estimated to be between 40x-49x.*[342]

Obviously, lower ADR numbers favor regulators' and manufacturers' position in providing drugs and hide the true weakness of the medications.

Perhaps a more dreadful aspect of this data collection and monitoring exercise has been the promotion of it as a modern substitute for experimental-based scientific research under the fanciful names, epidemiology, evidence-based regulation, data science, etc.

For example, the recent introduction of COVID-19 vaccines has been accepted based on theoretical models with arbitrary and irrelevant testing—mostly without toxicological assessments—arguing that toxicity or adverse effects could easily and effectively be observed in surveillance (ADR) monitoring.[343]

Furthermore, the biased nature of ADR monitoring is evident in the following example. Recently, CDC advised the public:

> *Stop using eye drops linked to drug-resistant bacteria outbreak...*

...based on the information it received, stating:

> *The Centers for Disease Control and Prevention*

342

https://amps.redunion.com.au/hubfs/Altman%20Report%20Version%209-8-22%20FINAL%20FINAL_%20(1).pdf)

343 https://bioanalyticx.com/the-fda-committees-review-of-pfizer-biontech-covid-19-vaccine-unscientific-false-and-deceitful/

> *on Wednesday night sent a health alert to physicians, saying the outbreak includes at least 55 people in 12 states. One died.*[344]

However, in the case of COVID-19 vaccines, thousands and thousands of adverse effects, with more than 30,000 reported deaths, have been recorded on the CDC ADR monitoring database (VAER) without concerns about adverse effects or suggestions for stopping or withdrawing vaccines.[345]

In fact, the opposite:

> *Yes. The two mRNA vaccines, Pfizer and Moderna, authorized by the U.S. Food and Drug Administration (FDA) and recommended by the Centers for Disease Control and Prevention (CDC), are very safe and very good at preventing serious or fatal cases of COVID-19. The risk of serious side effects associated with these vaccines is very small.*[346]

From the CDC site:

- *COVID-19 vaccines are safe and effective.*
- *Millions of people in the United States have received COVID-19 vaccines under the most intense safety monitoring in US history.*

> *COVID-19 vaccines are safe and effective.*

---

[344] https://www.latimes.com/science/story/2023-02-02/stop-using-eye-drops-linked-to-drug-resistant-bacteria-outbreak-cdc-says

[345] https://wonder.cdc.gov/vaers.html

[346] https://www.hopkinsmedicine.org/health/conditions-and-diseases/coronavirus/is-the-covid19-vaccine-safe

> *COVID-19 vaccines were evaluated in tens of thousands of participants in clinical trials. The vaccines met the Food and Drug Administration's (FDA's) rigorous scientific standards for safety, effectiveness, and manufacturing quality needed to support emergency use authorization (EUA).*[347]

The views mentioned above do not appear as views from a regulatory authority but from a vaccine manufacturer who likes to sell the products and downplay anything negative about their products.

So, just like the lack of science in medical/pharmaceutical claims and practices, monitoring the ADR effect is equally flawed and useless in providing accurate and unbiased information about medicines/vaccines and their adverse reactions.

The current promotion of vaccines touches an order higher level of deception by creating fear in the trusting public, not only the "sick," but perfectly healthy ones too. Unlike the drugs for treating illnesses, vaccines have been recommended and mandated for use in healthy people, i.e., vaccines as *health food*—arguably a significant extension of market size for drugs and vaccines.

Vaccine use has been promoted and mandated for the healthy with a rationale not only to protect them but their loved ones because the virus from the "unprotected" or unvaccinated can infect the "protected" ones or other vulnerable, e.g., the elderly. But, again, this is not monitoring or establishing drugs/vaccines' safety and efficacy but their marketing.

Note that there has been no evidence provided for the existence of the virus, let alone its transfer or spread from person to person. Interestingly, Pfizer acknowledges that no studies were conducted to establish if the virus transfers from person to

---

[347] https://www.cdc.gov/coronavirus/2019-ncov/vaccines/safety/safety-of-vaccines.html

person.[348]

At the same time, extensive and frequent use of testing has been implemented, such as for travel, employment, frequent and multiple testing with or without any relevant symptoms, etc. This resulted in a higher number of positives independent of vaccination status, just on a probability basis.

COVID-19 testing is based on the random detection of arbitrary RNA and s-protein components. It is assumed and hypothesized that testing positive indicated the presence of the virus or its infection.

Genetics, lifestyle, and environment are the biggest determinants of health reflected by normal variations in genetic pieces (so-called RNA or DNA fragments), which could easily be picked by PCR tests falsely claiming sickness (e.g., COVID-19 or any other). Hence, declaring a healthy population as sick warranting vaccination, as explained in Chapter 16 (Part 1).

However, the frequency of positive test results after vaccination, which should have indicated the deficiency of vaccines, has been twisted into claims of virus mutation. Hence more and/or different vaccine varieties—so-called boosters have been "developed" and promoted. It is hard to believe experts made such illogical and stupid claims that they could not show any specimen of the original virus but continue their hogwash by declaring and "identifying" its variants.

It establishes without a doubt a serious lack of understanding and practice of science in the medical field.

On top of that, medical and pharmaceutical experts developed and mandated vaccines against a non-existing virus or its illness. Worse, they assumed that the chemical (mRNA, a foreign molecule), when injected into the body, would not disturb the body's natural processes and chemistry and would be free from

---

[348] https://multimedia.europarl.europa.eu/en/webstreaming/special-committee-on-covid-19-pandemic_20221010-1430-COMMITTEE-COVI

producing significant adverse side effects. How naive.

They have been wrong here, too, as numerous ADRs, including deaths have been reported. But, rather than accepting them, they started hiding them by manipulating the reported ADR numbers. It is a sad story of the medical and pharmaceutical professions' false scientific claims.

Hence, it makes a strong case against monitoring the ADRs by authorities (such as CDC or FDA) and suggests that this practice be immediately stopped and, in the future, be managed by a truly independent third party, i.e., independent of the regulatory authorities and the drug manufacturers, more like a consumer-based one.

Links to Resources

WE GIVE THANKS to the website, patriots.win, who have kindly compiled a list of law firms that you may call if you believe your rights are being infringed upon by your employer, school, college, or any group that discriminates against you for not complying with their vaccine or mask mandates.[349]

The list includes:
1. Liberty Counsel https://lc.org/
2. Liberty Institute https://www.libertyinstitute.org/about/faq
3. Pacific Justice Institute
   https://www.secure.pacificjustice.org/site/SPageNavigator/contact_us.html
4. Advocates For Faith and Freedom
https://faith-freedom.com/
5. Alliance Defending Freedom https://adflegal.org/about-us
6. National Legal Foundation https://nationallegalfoundation.org/
7. Thomas More Law Center https://www.thomasmore.org/
8. Thomas More Society https://thomasmoresociety.org/
9. Christian Legal Society https://www.christianlegalsociety.org/
10. American Center for Law and Justice https://aclj.org/
11. Center for Law and Religious Freedom
https://www.clsreligiousfreedom.org/about-center
12. Christian Attorneys of America
https://christianattorneysofamerica.com/
13. Christian Law Association https://www.christianlaw.org/
14. National Association of Christian Lawmakers
https://christianlawmakers.com/
15. Pacific Legal Foundation https://pacificlegal.org/

---

[349] https://patriots.win/p/12kFnFBEiV/legal-resources-for-recourse-aga

Additional Legal Resources—Including Lawyers to Represent us and Treatment Options/Sources for HCQ and Ivermectin

50-state Update on Pending Employer-Mandated Vaccination Legislation (Husch Blackwell, LLP)
https://www.huschblackwell.com/newsandinsights/50-state-update-on-pending-legislation-pertainingto-employer-mandated-vaccinations

America's Frontline Doctors
https://americasfrontlinedoctors.org
https://americasfrontlinedoctors.org/legal
https://americasfrontlinedoctors.org/covid19

HCQ / Ivermectin Rx and Mask Exemptions
https://americasfrontlinedoctors.org/treatments/
Thomas Renz—Renz Law (Entire firm is dedicated to fighting for Medical Freedom. They work with Frontline Docs)
https://renz-law.com https://renz-law.com/our-medical-freedom-fight

Vaxx Injury Lawsuit—Complaint and Whistleblower Declaration
https://renz-law.com/45k-whistleblower-suit Right To Refuse Vaxx—Memorandum of Law KrisAnne Hall, JD
https://krisannehall.com/index.php/resources/articles/718-right-to-refuse-memorandum-of-law

Sidney Powell—Defending the Republic—Push Back against the Vaxx
https://defendingtherepublic.org/covid/

Employer Assumption of Liability & Religious Exemptions 'Form Examplars' for you to use or to consult with an attorney.

Assumption of Liability and Religious exemption (form for Protestants & Catholics)
Barnes Law (Robert Barnes) https://www.barneslawllp.com

Vaccine Mandate Protest Letter Example
https://vivabarneslaw.locals.com/post/689810/vaccine-mandate-employee-letter-example
https://vivabarneslaw.locals.com/upost/889925/vaccine-mandate-protest-letter
Colorado Catholic Bishops—template to for Vaxx religious exemption (8/5 Letter and Template)
https://cocatholicconference.org

Religious Exemption—GRANTED
https://greatawakening.win/p/12jvyovQwh/religious-exemption--granted/c

Federal Law Prohibits Mandates of Emergency Use COVID Vaccines, Tests, Masks—Resources to Inform Your School or Employer
https://msic69.wordpress.com/2021/07/29/federal-law-prohibits-mandates-of-emergency-use-covidvaccines-tests-masks-3-resources-you-can-use-to-inform-your-school-or-employer

Truth for Health Foundation: Legal Resources & Medical Censorship Defense Fund
https://www.truthforhealth.org/legal-resources

Vaxx Choice
https://www.vaxxchoice.com

File a Criminal Complaint
https://www.vaxxchoice.com/wp-content/uploads/2021/06/Combined-Criminal-Complaint-wInstructions-for-Filing-1.pdf
Refuse the Covid-19 Vaccine

https://www.vaxxchoice.com/wp-content/uploads/2021/07/Nuremberg-Notice-Form.pdf
Informed Consent Action Network (ICAN)
https://www.icandecide.org
Info on how to report Covic Vaxx injuries, legal / lawsuits, vaxx research, etc. They've already won a case against an employer. Email them at: freedom@icandecide.org.
ABA Journal article about Siri & Glimstad—working with ICAN to send out vaxx cease and desist letters
https://www.abajournal.com/news/article/this-law-firm-is-fighting-mandatory-covid-19-vaccines-withlegal-filings-and-warnings

Liberty Counsel Action—Resources, including Legal Help
https://lcaction.org

Vaccine Memo here: https://lcaction.org/vaccine
Andrew Torba—COVID Vaccine Religious Exemption
https://news.gab.com/2021/07/29/important-download-covid-vaccine-religious-exemption-documents
https://news.gab.com/tag/religious-exemption
Siri & Glimstad (NYC law firm) This firm is fighting mandatory COVID-19 vaccines with legal filings and warnings
https://www.sirillp.com

See the "Practice Area" of website for Vaccine Injury and Vaccine Exemptions information
https://www.abajournal.com/news/article/this-law-firm-is-fighting-mandatory-covid-19-vaccines-withlegal-filings-and-warnings

Curated Collection of the best mask studies and common sense arguments:
https://patriots.win/p/HEgcXddD/forcing-me-to-wear-political-sym

Another form to consider using when US employer is mandating jabs: https://pandemic.solari.com/form-for-employees-whose-employers-are-requiring-covid-19-injections
You Demand My Vax Papers? Well Then...Let's Go One Step Further https://patriots.win/p/12jcY1Oq1J/oh-you-demand-my-vax-papers-well/c
Association of American Physicians and Surgeons (AAPS) Legal Resources https://aapsonline.org/category/legal_matters

Open Letter from Physicians to Universities: Allow Students Back Without COVID Vaccine Mandate
https://aapsonline.org/open-letter-from-physicians-to-universities-reverse-covid-vaccine-mandates
18 Reasons I Won't Be Getting a COVID Vaccine (Source: TheDefender)
https://archive.is/CTmVV#selection-1073.0-1077.175

Rights and Freedoms: Covid-19
https://rightsfreedoms.wordpress.com
Federal Law Prohibits Mandates of COVID Vaccines, Tests, Masks—Resources to Inform Your School or Employer
https://rightsfreedoms.wordpress.com/2021/05/22/federal-law-prohibits-mandates-of-emergency-usecovid-vaccines-tests-masks-3-resources-you-can-use-to-inform-your-school-or-employer
Form (and lots of related resources) for Employees Whose Employers Are Requiring Covid-19 Injections (Solari Report)
https://pandemic.solari.com/form-for-employees-whose-employers-are-requiring-covid-19-injections

Children's Health Defense (EUROPE)
https://childrenshealthdefense.org
https://childrenshealthdefense.org/legal/legal-resources

Notices for EUA Masks / EUA Testing / EUA Vaccines. Health Freedom Defense Fund (See Resources section)

https://healthfreedomdefense.org
Los Angeles Unified School District Abandons Mandatory Vaccination Due to Lawsuit
https://healthfreedomdefense.org/2021/08/los-angeles-unified-school-district-abandons-mandatoryvaccination-due-to-lawsuit
18 Reasons I Won't Be Getting a COVID Vaccine
https://childrenshealthdefense.org/defender/reasons-not-getting-covid-vaccine

How to refuse the jab without refusing the jab—ask for more information. (They CANNOT provide that information but you've NOT refused.)
https://greatawakening.win/p/12jd0GC8LQ/how-to-refuse-the-jab-without-re/c
https://gab.com/M_johnson1016/posts/106695311646239538

Catholic Medical Association Joint Statement: Vaccines and Conscience Protection
https://www.cathmed.org/vaccines-and-conscience-protection
https://www.cathmed.org/programs-resources/cma-resources/coronavirus-resources

Canada—Ontario Civil Liberties Association—Letter
https://dailyexpose.co.uk/2021/08/04/a-letter-to-the-unvaccinated
https://archive.md/UnULV

Federal Employees
https://archive

---

[1] Source: The Association of Faculties of Medicine of Canada (AFMC), 2007, *The Interacting Triad of Causal Factors*. AFMC Public Health Educators' Network https://phprimer.afmc.ca/en/part-i/chapter-2/ (accessed September 2022). License: Creative Commons BY-NC-SA

Milton Keynes UK
Ingram Content Group UK Ltd.
UKHW011344150124
436064UK00004B/418